Advanced Analytics with Transact-SQL

Exploring Hidden Patterns and Rules in Your Data

Dejan Sarka

Apress®

Advanced Analytics with Transact-SQL: Exploring Hidden Patterns and Rules in Your Data

Dejan Sarka
Ljubjana, Slovenia

ISBN-13 (pbk): 978-1-4842-7172-8   ISBN-13 (electronic): 978-1-4842-7173-5
https://doi.org/10.1007/978-1-4842-7173-5

Managing Director, Apress Media LLC: Welmoed Spahr
Acquisitions Editor: Jonathan Gennick
Development Editor: Laura Berendson
Coordinating Editor: Jill Balzano

Cover image designed by Freepik (www.freepik.com)

Distributed to the book trade worldwide by Springer Science+Business Media LLC, 1 New York Plaza, Suite 4600, New York, NY 10004. Phone 1-800-SPRINGER, fax (201) 348-4505, e-mail orders-ny@springer-sbm. com, or visit www.springeronline.com. Apress Media, LLC is a California LLC and the sole member (owner) is Springer Science + Business Media Finance Inc (SSBM Finance Inc). SSBM Finance Inc is a **Delaware** corporation.

For information on translations, please e-mail booktranslations@springernature.com; for reprint, paperback, or audio rights, please e-mail bookpermissions@springernature.com.

Apress titles may be purchased in bulk for academic, corporate, or promotional use. eBook versions and licenses are also available for most titles. For more information, reference our Print and eBook Bulk Sales web page at http://www.apress.com/bulk-sales.

Any source code or other supplementary material referenced by the author in this book is available to readers on GitHub via the book's product page, located at www.apress.com/9781484271728. For more detailed information, please visit http://www.apress.com/source-code.

Printed on acid-free paper

Table of Contents

About the Author

Dejan Sarka, MCT and Data Platform MVP, is an independent trainer and consultant who focuses on developing database and business intelligence applications, with more than 30 years of experience in this field. Besides projects, he spends about half of his time on training and mentoring. He is the founder of the Slovenian SQL Server and .NET Users Group. Sarka is the author or co-author of 19 books about databases and SQL Server. He has developed many courses and seminars for Microsoft, RADACAD, SolidQ, and Pluralsight.

About the Technical Reviewer

Ed Pollack has more than 20 years of experience in database and systems administration, developing a passion for performance optimization, database design, and wacky analytics. He has spoken at many user groups, data conferences, and summits. This led him to organize SQL Saturday Albany, which has become an annual event for New York's Capital Region. Sharing these experiences with the community is a passion. In his free time, Ed enjoys video games, backpacking, traveling, and cooking exceptionally spicy foods.

Acknowledgments

Many people helped me with this book, directly or indirectly. I am not mentioning the names explicitly because I could unintentionally omit some of them. However, I am pretty sure they are going to recognize themselves in the following text.

Thank you to my family, who understood that I could not spend so much time with them while I am writing the book.

Thank you to my friends, who were always encouraging me to finish the work.

Thank you to all the great people that work for or are engaged with Apress. Without your constant support, this book would probably never be finished.

Thank you to the reviewers. A well-reviewed book is as important as the writing itself. Thank you very much for your work.

Introduction

If you want to learn how to get information from your data with Transact-SQL, or the T-SQL language, this book is for you. It teaches you how to calculate statistical measures from descriptive statistics, including centers, spreads, skewness, and the kurtosis of a distribution, find the associations between pairs of variables, including calculating the linear regression formula, calculate the confidence level with definite integration, find the amount of information in your variables, and also do some machine learning or data science analysis, including predictive modeling and text mining.

The T-SQL language is in the latest editions of SQL Server, Azure SQL Database, and Azure Synapse Analytics. It has so many business intelligence (BI) improvements that it might become your primary analytic database system. Many database developers and administrators are already proficient with T-SQL. Occasionally they need to analyze the data with statistical or data science methods, but they do not want to or have time to learn a completely new language for these tasks. In addition, they need to analyze huge amounts of data, where specialized languages like R and Python might not be fast enough. SQL Server has been optimized for work with big datasets for decades.

To get the maximum out of these language constructs, you need to learn how to use them properly. This in-depth book shows extremely efficient statistical queries that use the window functions and are optimized through algorithms that use mathematical knowledge and creativity. The formulas and usage of those statistical procedures are explained as well.

Any serious analysis starts with data preparation. This book introduces some common data preparation tasks and shows how to implement them in T-SQL.

No analysis is good without good data quality. The book introduces data quality issues and shows how you can check for completeness and accuracy with T-SQL and measure improvements in data quality over time. It also shows how you can optimize queries with temporal data; for example, when you search for overlapping intervals. More advanced time-oriented information includes hazard and survival analysis.

Next, the book turns to data science. Some advanced algorithms can be implemented in T-SQL. You learn about the market basket analysis with association rules using different measures like support and confidence, and sequential market basket analysis when there is a sequence in the basket. Then the book shows how to develop predictive models with a mixture of k-nearest neighbor and decision tree algorithms and Bayesian inference analysis.

Analyzing text, or text mining, is a popular topic. You can do a lot of text mining in pure T-SQL, and SQL Server can become a text mining engine. The book explains how to analyze text in multiple natural languages with pure T-SQL and features from full-text search (FTS).

In short, this book teaches you how to use T-SQL for

- statistical analysis

- data science methods

- text mining

Who Should Read This Book

Advanced Analytics with Transact-SQL is for database developers and database administrators who want to take their T-SQL programming skills to the max. It is for those who want to efficiently analyze huge amounts of data by using their existing knowledge of the T-SQL language. It is also for those who want to improve querying by learning new and original optimization techniques.

Assumptions

This book assumes that the reader already has good knowledge of the Transact-SQL language. A few years of coding experience is very welcome. A basic grasp of performance tuning and query optimization can help you better understand how the code works.

The Organization of This Book

There are eight chapters in this book, which are logically structured in four parts, each part with two chapters. The following is a brief description of the chapters.

Part I: Statistics Most advanced analytics starts with good old statistics. Sometimes statistical analysis might already provide the needed information, and sometimes statistics is only used in an overview of the data.

Chapter 1: Descriptive Statistics With descriptive statistics, the analyst gets an understanding of the distribution of a variable. One can analyze either continuous or discrete variables. Depending on the variable type, the analyst must choose the appropriate statistical measures.

Chapter 2: Associations Between Pairs of Variables When measuring associations between pairs of variables, there are three possibilities: both variables are continuous, both are discrete, or one is continuous and the other one is discrete. Based on the type of the variables, different measures of associations can be calculated. To calculate the statistical significance of associations, the calculation of the definite integrals is needed.

Part II: Data Quality and Preparation Before doing advanced analyses, it is crucial to understand the quality of the input data. A lot of additional work with appropriate data preparation is usually a big part of an analytics project in real life.

Chapter 3: Data Preparation There is no end to data preparation tasks. Some of the most common tasks include converting strings to numerical variables and discretizing continuous variables. Missing values are typical in analytical projects. Many times, derived variables help explain the values of a target variable more than the original input.

Chapter 4: Data Quality Garbage in, garbage out is a very old rule. Before doing advanced analyses, it is always recommendable to check for the data quality. Measuring improvements in data quality over time can help with understanding the factors that influence it.

Part III: Dealing with Time Queries that deal with time-oriented data are very frequent in analytical systems. Beyond a simple comparison of data in different time periods, much more complex problems and analyses can arise.

Chapter 5: Time-Oriented Data Understanding what kind of temporal data can appear in a database is very important. Some types of queries that deal with temporal data are hard to optimize. Data preparation of time series data has some own rules as well.

Chapter 6: Time-Oriented Analyses How long is a customer faithful to the supplier or the subscribed services and service provider? Which are the most hazardous days for losing a customer? What will be the sales amount in the next few periods? This chapter shows how to answer these questions with T-SQL.

Part IV: Data Science Some of the most advanced algorithms for analyzing data are many times mentioned with the term data science. Expressions as data mining, machine learning, and text mining are also very popular.

Chapter 7: Data Mining Every online or retail shop wants to know which products customers tend to buy together. Predicting a target discrete or continuous variable with few input variables is important in practically every type of business. This chapter introduces some of the most popular algorithms implemented with T-SQL.

Chapter 8: Text Mining The last chapter of the book introduced text mining with T-SQL. Text mining can include semantic search, term extraction, quantitative analysis of words and characters, and more. Data mining algorithms like association rules can also be used to understand analyzed text better.

System Requirements

You need the following software to run the code samples in this book.

- Microsoft SQL Server 2019 Developer of Enterprise edition, which is at `www.microsoft.com/en-us/sql-server/sql-server-2019`.

- Azure SQL Server Managed Instance, which is at `https://azure.microsoft.com/en-us/services/azure-sql/sql-managed-instance/`.

- Most of the code should run on the Azure SQL Database, which is at `https://azure.microsoft.com/en-us/services/sql-database/`.

- If you would like to try the code on unlimited resources, you can use Azure Synapse Analytics at `https://azure.microsoft.com/en-us/services/synapse-analytics/`.

- SQL Server Management Studio is the default client tool. You can download it for free at `https://docs.microsoft.com/en-us/sql/ssms/download-sql-server-management-studio-ssms?view=sql-server-ver15`.

- Another free client tool is the Azure Data Studio at `https://docs.microsoft.com/en-us/sql/azure-data-studio/download-azure-data-studio?view=sql-server-ver15`.

- For demo data, you can find the AdventureWorks sample databases at https://docs.microsoft.com/en-us/sql/samples/ adventureworks-install-configure?view=sql-server- ver15&tabs=ssms.

- Some demo data comes from R. I explain how to get it and show the R code for loading the demo data in SQL Server.

- You can download all the code used for the companion content in this book at https://github.com/Apress/adv-analytics-w- transact-sql.

Naming Conventions

When I create tables in SQL Server, I start with the column(s) that form the primary key, and I use Pascal case (e.g., FirstName) for the physical columns. For computed, typically aggregated columns from a query, I tend to use camel case (e.g., avgAmount). However, the book deals with the data from many sources. Demo data provided from Microsoft demo databases is not enough for all my examples. Two demo tables come from R. In R, the naming convention is not strict. I had a choice on how to proceed.

I decided to go with the original names when data comes from R, so the names of the columns in the table are all lowercase (e.g., carbrand). However, Microsoft demo data is far from perfect as well. Many dynamic management objects return all lowercase objects or even reserved keywords as the names of the columns. For example, in Chapter 8, I use two tabular functions provided by Microsoft, which return two columns with names [KEY] and [RANK]. Both in uppercase, and both are even reserved words in SQL, so they need to be enclosed in brackets. This is why sometimes the reader might get the impression that the naming convention is not good. I made an arbitrary decision, which I hope was the best in this situation.

PART I

Statistics

CHAPTER 1

Descriptive Statistics

Descriptive statistics summarize or quantitatively describe variables from a dataset. In a SQL Server table, a *dataset* is a set of the rows, or a *rowset*, that comes from a SQL Server table, view, or tabular expression. A *variable* is stored in a column of the rowset. In statistics, a variable is frequently called a *feature*.

When you analyze a variable, you first want to understand the *distribution* of its values. You can get a better understanding through graphical representation and descriptive statistics. Both are important. For most people, a graphical representation is easier to understand. However, with descriptive statistics, where you get information through numbers, it is simpler to analyze a lot of variables and compare their aggregated values; for example, their means and variability. You can always order numbers and quickly notice which variable has a higher mean, median, or other measure.

Transact-SQL is not very useful for graphing. Therefore, I focus on calculating descriptive statistics measures. I also include a few graphs, which I created with Power BI.

Variable Types

Before I calculate the summary values, I need to introduce the types of variables. Different types of variables require different calculations. The most basic division of the Variables are basically divided into two groups: discrete and continuous.

Discrete variables can only take a value from a limited pool. For example, there are only seven different or distinct values for the days of the week. Discrete variables can be further divided into two groups: nominal and ordinal.

If a value does not have a quantitative value (e.g., a label for a group), it is a nominal variable. For example, a variable that describes marital status could have three possible values: single, married, or divorced.

Discrete variables could also have an intrinsic order, which are called *ordinal* variables. If the values are represented as numbers, it is easy to notice the order. For example, evaluating a product purchased on a website could be expressed with numbers

© Dejan Sarka 2021
D. Sarka, *Advanced Analytics with Transact-SQL*, https://doi.org/10.1007/978-1-4842-7173-5_1

from 1 to 7, where a higher number means greater satisfaction with the product. If the values of a variable are represented with strings, it is sometimes harder to notice the order. For example, education could be represented with strings, like high school degree, graduate degree, and so forth. You probably don't want to sort the values alphabetically because there is an order hidden in the values. With education, the order is defined through the years of schooling needed to get the degree.

If a discrete variable can take only two distinct values, it is a dichotomous variable called an *indicator*, a *flag*, or a *binary* variable. If the variable can only take a single value, it is a *constant*. Constants are not useful for analysis; there is no information in a constant. After all, variables are called *variables* because they introduce some variability.

Continuous variables can take a value from an unlimited, uncountable set of possible values. They are represented with integral or decimal numbers. They can be further divided into two classes: intervals or numerics (or true numerics).

Intervals are limited on the lower side, the upper side, or both sides. For example, temperature is an interval, limited with absolute zero on the lower side. On the other hand, true *numerics* have no limits on any side. For example, cashflow can be positive, negative, or zero.

It is not always completely clear if a variable is discrete or continuous. For example, the number of cars owned is an integer and can take any value between zero and infinite. You can use such variables in both ways—as discrete, when needed, or as continuous. For example, the naïve Bayes algorithm, which is explained in Chapter 7, uses only discrete variables so that you can treat the number of cars owned variable as discrete. But the linear regression algorithm, which is explained in the same chapter, uses only continuous variables, and you can treat the same variable as continuous.

Demo Data

I use a couple of demo datasets for the demos in this book. In this chapter, I use the *mtcars* demo dataset that comes from the R language; mtcars is an acronym for *MotorTrend* Car Road Tests. The dataset includes 32 *cases*, or rows, originally with 11 variables. For demo purposes, I add a few calculated variables. The data comes from a 1974 *MotorTrend* magazine and includes design and performance aspects for 32 cars, all 1973 and 1974 models. You can learn more about this dataset at `www.rdocumentation.org/packages/datasets/versions/3.6.2/topics/mtcars`.

I introduce variables when needed.

From SQL Server 2016, it is easy to execute R code inside SQL Server Database Engine. You can learn more about machine learning inside SQL Server with R or the Python language in official Microsoft documentation. A good introduction is at https://docs.microsoft.com/en-us/sql/machine-learning/sql-server-machine-learning-services?view=sql-server-ver15. Since this book is about T-SQL and not R, I will not spend more time explaining the R part of the code. I introduce the code that I used to import the mtcars dataset, with some additional calculated columns, in a SQL Server table.

First, you need to enable external scripts execution in SQL Server.

```
-- Configure SQL Server to enable external scripts
USE master;
EXEC sys.sp_configure 'show advanced options', 1;
RECONFIGURE
EXEC sys.sp_configure 'external scripts enabled', 1;
RECONFIGURE;
GO
```

I created a new table in the AdventureWorksDW2017 demo database, which is a Microsoft-provided demo database. I use the data from this database later in this book as well. You can find the AdventureWorks sample databases at https://docs.microsoft.com/en-us/sql/samples/adventureworks-install-configure?view=sql-server-ver15&tabs=ssms. For now, I won't spend more time on the content of this database. I just needed a database to create a table in, and because I use this database later, it seems like the best place for my first table with demo data. Listing 1-1 shows the T-SQL code for creating the demo table.

Listing 1-1. Creating the Demo Table

```
-- Create a new table in the AWDW database
USE AdventureWorksDW2017;
DROP TABLE IF EXISTS dbo.mtcars;
CREATE TABLE dbo.mtcars
(
 mpg numeric(8,2),
 cyl int,
 disp numeric(8,2),
```

```
hp int,
drat numeric(8,2),
wt numeric(8,3),
qsec numeric(8,2),
vs int,
am int,
gear int,
carb int,
l100km numeric(8,2),
dispcc numeric(8,2),
kw numeric(8,2),
weightkg numeric(8,2),
transmission nvarchar(10),
engine nvarchar(10),
hpdescription nvarchar(10),
carbrand nvarchar(20) PRIMARY KEY
)
GO
```

I want to discuss the naming conventions in this book. When I create tables in SQL Server, I start with the column(s) that form the primary key and use pascal case (e.g., FirstName) for the physical columns. For computed columns, typically aggregated columns from a query, I tend to use camel case (e.g., avgAmount). However, the book deals with data from many sources. Demo data provided from Microsoft demo databases is not enough for all of my examples. Two demo tables come from R. In R, the naming convention is not strict. I had a choice to make on how to proceed. I decided to go with the original names when data comes from R, so the names of the columns in the table in Listing 1-1 are all lowercase (e.g., carbrand).

Note   Microsoft demo data is far from perfect. Many dynamic management objects return all lowercase objects or reserved keywords as the names of the columns. For example, in Chapter 8, I use two tabular functions by Microsoft that return two columns named [KEY] and [RANK]. Both are uppercase reserved words in SQL, so they need to be enclosed in brackets.

Now let's use the `sys.sp_execute_external_script` system stored procedure to execute the R code. Listing 1-2 shows how to execute the `INSERT...EXECUTE` T-SQL statement to get the R dataset in a SQL Server table.

Listing 1-2. Inserting R Data in the SQL Server Demo Table

```
-- Insert the mtcars dataset
INSERT INTO dbo.mtcars
EXECUTE sys.sp_execute_external_script
 @language=N'R',
 @script = N'
data("mtcars")
mtcars$l100km = round(235.214583 / mtcars$mpg, 2)
mtcars$dispcc = round(mtcars$disp * 16.38706, 2)
mtcars$kw = round(mtcars$hp * 0.7457, 2)
mtcars$weightkg = round(mtcars$wt * 1000 * 0.453592, 2)
mtcars$transmission = ifelse(mtcars$am == 0,
                             "Automatic", "Manual")
mtcars$engine = ifelse(mtcars$vs == 0,
                       "V-shape", "Straight")
mtcars$hpdescription =
  factor(ifelse(mtcars$hp > 175, "Strong",
               ifelse(mtcars$hp < 100, "Weak", "Medium")),
         order = TRUE,
         levels = c("Weak", "Medium", "Strong"))
mtcars$carbrand = row.names(mtcars)
 ',
 @output_data_1_name = N'mtcars';
GO
```

You can check if the demo data successfully imported with a simple `SELECT` statement.

```
SELECT *
FROM dbo.mtcars;
```

When the demo data is loaded, let's start analyzing it.

Frequency Distribution of Discrete Variables

You usually represent the distribution of a discrete variable with frequency distribution or *frequencies*. In the simplest example, you can calculate only the values' count. You can also express these value counts as percentages of the total number of rows or cases.

Frequencies of Nominals

The following is a simple example of calculating the counts and percentages for the *transmission* variable, which shows the transmission type.

```
-- Simple, nominals
SELECT c.transmission,
 COUNT(c.transmission) AS AbsFreq,
 CAST(ROUND(100. * (COUNT(c.transmission)) /
      (SELECT COUNT(*) FROM mtcars), 0) AS int) AS AbsPerc
FROM dbo.mtcars AS c
GROUP BY c.transmission;
```

The following is the result.

```
transmission AbsFreq     AbsPerc
------------ ----------- -----------

Automatic    19          59
Manual       13          41
```

I used a simple GROUP BY clause of the SELECT statement and the COUNT() aggregate function. Graphically, you can represent the distribution with vertical or horizontal bar charts. Figure 1-1 shows the bar charts for three variables from the mtcars dataset, created with Power BI.

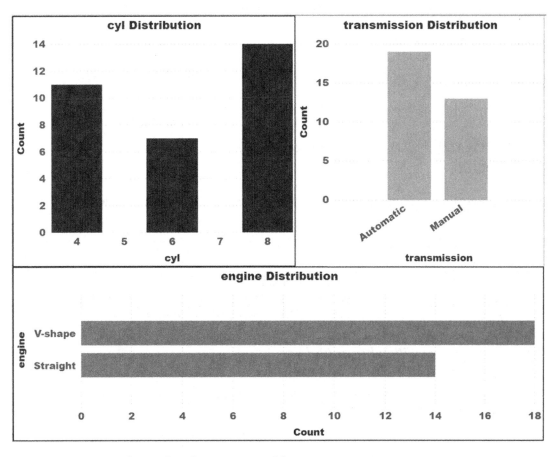

Figure 1-1. *Bar charts for discrete variables*

You can see the distribution of the transmission, engine, and cyl variables. The cyl variable is represented with the numbers 4, 6, and 8, which represent the number of engine cylinders. Can you create a bar chart with T-SQL? You can use the percentage number as a parameter to the REPLICATE() function and mimic the horizontal bar chart, or a horizontal histogram, as the following code shows.

```
WITH freqCTE AS
(
SELECT c.transmission,
 COUNT(c.transmission) AS AbsFreq,
 CAST(ROUND(100. * (COUNT(c.transmission)) /
      (SELECT COUNT(*) FROM mtcars), 0) AS int) AS AbsPerc
```

```
FROM dbo.mtcars AS c
GROUP BY c.transmission
)
SELECT transmission,
 AbsFreq,
 AbsPerc,
 CAST(REPLICATE('*', AbsPerc) AS varchar(50)) AS Histogram
FROM freqCTE;
```

I used a common table expression to enclose the first query, which calculated the counts and the percentages, and then added the horizontal bars in the outer query. Figure 1-2 shows the result.

	transmission	AbsFreq	AbsPerc	Histogram
1	Automatic	19	59	***
2	Manual	13	41	***

Figure 1-2. *Counts with a horizontal bar*

For nominal variables, this is usually all that you calculate. For ordinals, you can also calculate running totals.

Frequencies of Ordinals

Ordinals have intrinsic order. When you sort the values in the correct order, it makes sense to also calculate the running totals. What is the total count of cases up to some specific value? What is the running total of percentages? You can use the T_SQL *window aggregate functions* to calculate the running totals. Listing 1-3 shows the calculation for the cyl variable.

Listing 1-3. Frequencies of an Ordinal Variable

```
-- Ordinals - simple with numerics
WITH frequency AS
(
SELECT v.cyl,
 COUNT(v.cyl) AS AbsFreq,
 CAST(ROUND(100. * (COUNT(v.cyl)) /
```

```
       (SELECT COUNT(*) FROM dbo.mtcars), 0) AS int) AS AbsPerc
FROM dbo.mtcars AS v
GROUP BY v.cyl
)
SELECT cyl,
 AbsFreq,
 SUM(AbsFreq)
  OVER(ORDER BY cyl
       ROWS BETWEEN UNBOUNDED PRECEDING
       AND CURRENT ROW) AS CumFreq,
 AbsPerc,
 SUM(AbsPerc)
  OVER(ORDER BY cyl
       ROWS BETWEEN UNBOUNDED PRECEDING
       AND CURRENT ROW) AS CumPerc,
 CAST(REPLICATE('*', AbsPerc) AS varchar(50)) AS Histogram
FROM frequency
ORDER BY cyl;
```

The query returns the result shown in Figure 1-3.

	cyl	AbsFreq	CumFreq	AbsPerc	CumPerc	Histogram
1	4	11	11	34	34	**********************************
2	6	7	18	22	56	**********************
3	8	14	32	44	100	**

Figure 1-3. *Frequencies of an ordinal variable*

Note   If you are not familiar with the T-SQL window functions and the OVER() clause, please refer to the official SQL Server documentation at `https://docs. microsoft.com/en-us/sql/t-sql/queries/select-over-clause-transact-sql?view=sql-server-ver15`.

Ordering by the cyl variable was simple because the values are represented with integral numbers, and the order is automatically correct. But if an ordinal is represented with strings, you need to be careful with the proper order. You probably do not want to use alphabetical order.

For a demo, I created (already in the R code) a hpdescription derived variable (originally stored in the hp continuous variable), which shows engine horsepower in three classes: weak, medium, and strong. The following query incorrectly returns the result in alphabetical order.

```
-- Ordinals - incorrect order with strings
WITH frequency AS
(
SELECT v.hpdescription,
 COUNT(v.hpdescription) AS AbsFreq,
 CAST(ROUND(100. * (COUNT(v.hpdescription)) /
      (SELECT COUNT(*) FROM dbo.mtcars), 0) AS int) AS AbsPerc
FROM dbo.mtcars AS v
GROUP BY v.hpdescription
)
SELECT hpdescription,
 AbsFreq,
 SUM(AbsFreq)
  OVER(ORDER BY hpdescription
       ROWS BETWEEN UNBOUNDED PRECEDING
       AND CURRENT ROW) AS CumFreq,
 AbsPerc,
 SUM(AbsPerc)
  OVER(ORDER BY hpdescription
       ROWS BETWEEN UNBOUNDED PRECEDING
       AND CURRENT ROW) AS CumPerc,
 CAST(REPLICATE('*', AbsPerc) AS varchar(50)) AS Histogram
FROM frequency
ORDER BY hpdescription;
```

The results of this query are shown in Figure 1-4.

	hpdescription	AbsFreq	CumFreq	AbsPerc	CumPerc	Histogram
1	Medium	13	13	41	41	***
2	Strong	10	23	31	72	*******************************
3	Weak	9	32	28	100	****************************

Figure 1-4. *Frequencies of the hpdescription variable with incorrect order*

You can use the CASE T-SQL expression to change the strings and include proper ordering with numbers at the beginning of the string. Listing 1-4 shows the calculation of the frequencies of a string ordinal with proper ordering.

Listing 1-4. Frequencies of an Ordinal with Proper Ordering

```
-- Ordinals - correct order
WITH frequency AS
(
SELECT
 CASE v.hpdescription
         WHEN N'Weak' THEN N'1 - Weak'
         WHEN N'Medium' THEN N'2 - Medium'
         WHEN N'Strong' THEN N'3 - Strong'
      END AS hpdescriptionord,
 COUNT(v.hpdescription) AS AbsFreq,
 CAST(ROUND(100. * (COUNT(v.hpdescription)) /
      (SELECT COUNT(*) FROM dbo.mtcars), 0) AS int) AS AbsPerc
FROM dbo.mtcars AS v
GROUP BY v.hpdescription
)
SELECT hpdescriptionord,
 AbsFreq,
 SUM(AbsFreq)
  OVER(ORDER BY hpdescriptionord
      ROWS BETWEEN UNBOUNDED PRECEDING
      AND CURRENT ROW) AS CumFreq,
 AbsPerc,
 SUM(AbsPerc)
  OVER(ORDER BY hpdescriptionord
      ROWS BETWEEN UNBOUNDED PRECEDING
      AND CURRENT ROW) AS CumPerc,
 CAST(REPLICATE('*', AbsPerc) AS varchar(50)) AS Histogram
FROM frequency
ORDER BY hpdescriptionord;
```

Figure 1-5 shows the result of the query from Listing 1-4.

	hpdescriptionord	AbsFreq	CumFreq	AbsPerc	CumPerc	Histogram
1	1 - Weak	9	9	28	28	****************************
2	2 - Medium	13	22	41	69	**
3	3 - Strong	10	32	31	100	*******************************

Figure 1-5. *Frequencies of the hpdescription ordinal variable*

With frequencies, I covered discrete variables. Now let's calculate some descriptive statistics for continuous variables.

Descriptive Statistics for Continuous Variables

You can calculate many statistical values for the distribution of a continuous variable. Next, I show you the calculation for the centers of distribution, spread, skewness, and "tailedness." I also explain the mathematical formulas for calculation and the meaning of the measures. These measures help describe the distribution of a continuous variable without graphs.

Centers of a Distribution

The most known and the most abused statistical measure is the *mean* or the average of a *variable*. How many times have you heard or read about "the average ..."? Many times, this expression makes no sense, although it looks smart to use it. Let's discuss an example.

Take a group of random people in a bar. For the sake of the example, let's say they are all local people from the country where the bar is located. You want to estimate the wealth of these people.

The mean value is also called the *expected* value. It is used as the *estimator* for the target variable, in this case, wealth. It all depends on how you calculate the mean. You can ask every person in the group her or his income and then calculate the group's mean. This is the *sample* mean.

Your group is a sample of the broader population. You could also calculate the mean for the whole country. This would be the *population* mean. The population mean is a good estimator for the group. However, the sample mean could be very far from the actual wealth of the majority of people in the group. Imagine that there are 20 people in the group, including one extremely rich person worth more than $20 billion. The sample mean would be more than a billion dollars, which seems like a group of billionaires are in the bar. This could be far from the truth.

Extreme values, especially if they are rare, are called *outliers*. Outliers can have a big impact on the mean value. This is clear from the formula for the mean.

$$\mu = \frac{1}{n} * \sum_{i=1}^{n} v_i$$

Each value, v_i, is part of the calculation of the mean, μ. A value of 100 adds a hundred times more to the mean than the value of 1. The mean of the sample is rarely useful if it is the only value you are measuring. The calculation of the mean involves every value on the first degree. That is why the mean is also called the *first population moment*.

Apparently, we need other measures. A very useful measure is the median. First, you order the rows (or the cases) by the target variable. Then, you split the population into two halves, or two tiles. The median is the value in the middle. If you have an odd number of rows, it is a single value because there is exactly one row in the middle. If you have an even number of rows, there are two rows in the middle. You can calculate the median as the lower of these two values, which is then the lower median.

Similarly, the upper median is the higher of the two values in the middle. I prefer to use the average of the two values in the middle, called the *median*. You can also split the cases into more than two tiles. If you split it into four tiles, the tiles are called *quartiles*. The median is the value on the second quartile, sometimes marked as Q_2.

In T-SQL, you can use the PERCENTILE_CONT() window analytic function to calculate the median and the PERCENTILE_DISC() function to calculate the lower median. The following code shows how these two functions work on a small sample dataset that consists of values 1, 2, 3, and 4.

```
-- Difference between PERCENTILE_CONT() and PERCENTILE_DISC()
SELECT DISTINCT        -- can also use TOP (1)
 PERCENTILE_DISC(0.5) WITHIN GROUP
  (ORDER BY val) OVER () AS MedianDisc,
 PERCENTILE_CONT(0.5) WITHIN GROUP
  (ORDER BY val) OVER () AS MedianCont
FROM (VALUES (1), (2), (3), (4)) AS a(val);
```

Here are the results.

```
MedianDisc  MedianCont
----------- ----------------------
2           2.5
```

If the median is much lower than the mean, it means you have some extreme values or a long tail on the upper side or on the right side of the numbers line. If the median is higher than the mean, you have a tail on the left side; the distribution is skewed to the left. By comparing the median and the mean, you can infer something about the shape of the distribution. You calculate the mean with the AVG() T-SQL function.

Listing 1-5 shows the calculation for the mean and the median for the hp and weightkg variables. I use two common expressions to calculate the two medians. In the outer query, two SELECT statements combine their results in a single result set with the UNION operator. Although the query looks unnecessarily complex, I prefer to do it this way to have a single result, where it is simple to compare the two variables' values.

Listing 1-5. Calculation of the Median and the Mean

```
-- Mean and median
-- Weight and hp
WITH medianW AS
(
SELECT N'weight' AS variable, weightkg,
 PERCENTILE_CONT(0.5) WITHIN GROUP
  (ORDER BY weightkg) OVER () AS median
FROM dbo.mtcars
),
medianH AS
(
SELECT N'hp' AS variable, hp,
 PERCENTILE_CONT(0.5) WITHIN GROUP
  (ORDER BY hp) OVER () AS median
FROM dbo.mtcars
)
SELECT
 MIN(variable) AS variable,
```

```
 AVG(weightkg) AS average,
 MIN(median) AS median
FROM medianW
UNION
SELECT
 MIN(variable) AS variable,
 AVG(hp) AS average,
 MIN(median) AS median
FROM medianH
ORDER BY variable;
GO
```

Figure 1-6 shows the result.

	variable	average	median
1	hp	146.000000	123
2	weight	1459.318750	1508.195

Figure 1-6. *Average and median for two variables*

In Figure 1-6, the hp variable has an average greater than the median. Therefore, this variable has some tail on the right side. However, the weight variable has values close together; although the median is slightly greater than the mean, the difference is small. Therefore, there is no significant tail on any of the sides.

To get rid of the extreme values, you can also calculate the *trimmed mean.* You exclude a few (e.g., three) of the highest and lowest values and calculate the mean from the rest of the values, like the following code shows.

```
WITH rownumbers AS
(
SELECT hp,
 ROW_NUMBER() OVER(ORDER BY hp ASC) AS rna,
 ROW_NUMBER() OVER(ORDER BY hp DESC) AS rnd
FROM dbo.mtcars
)
SELECT AVG(hp) AS trimmedMean
FROM rownumbers
WHERE rna > 3 AND rnd > 3;
```

The trimmed mean for the hp variable, returned by the previous query, is 141.

Another measure that helps with understanding the shape of a variable's distribution is the *mode*. The mode is the most frequent or the most popular value. A variable can have more than one mode. This means that a variable has *multimodal* distribution. Figure 1-7 shows the distribution of the hp variable. The variable has a peak at three different values; it has three modes.

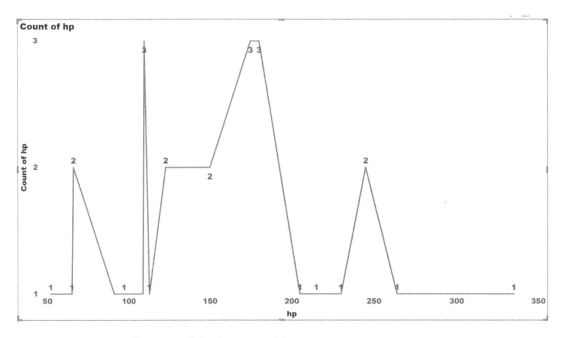

Figure 1-7. *Distribution of the hp variable*

You can calculate the mode with the TOP T-SQL operator. To catch all the modes, you can use the TOP 1 WITH TIES clause, like in Listing 1-6.

Listing 1-6. Calculating the Mode

```
-- Mode
-- Single mode
SELECT TOP (1) WITH TIES weightkg, COUNT(*) AS n
FROM dbo.mtcars
GROUP BY weightkg
ORDER BY COUNT(*) DESC;
-- Multimodal distribution
```

```
SELECT TOP (1) WITH TIES hp, COUNT(*) AS n
FROM dbo.mtcars
GROUP BY hp
ORDER BY COUNT(*) DESC;
GO
```

Here are the results.

```
weightkg                                       n
-------------------------------------- -----------
1560.36                                        3

hp          n
----------- -----------
110         3
175         3
180         3
```

The hp variable is multimodal, while the weightkg variable has a single peak and only a single mode.

Because of the difference between the mean and the median of the hp variable, you already know that this variable's values are not close around the mean. There is some spread in the distribution.

Measuring the Spread

As with centers, you can calculate multiple measures for the spread of a distribution. The simplest measure is the *range*. You calculate it by subtracting the minimum value from the maximum value, as the following formula shows.

$$R = v_{max} - v_{min}$$

The following code calculates the range for two variables.

```
-- Range
SELECT MAX(weightkg) - MIN(weightkg) AS rangeWeight,
 MAX(hp) - MIN(hp) AS rangeHp
FROM dbo.mtcars;
```

And here are the results.

```
rangeWeight    rangeHp
-------------  -----------
1774.00        283
```

A range gives a very limited amount of information, especially if there are outliers among the variable values. Usually, the *interquartile range* is a much more useful measure. The interquartile range, or IQR, is calculated by taking the value on the third quartile and subtracting the value of the first quartile. The formula is simple.

$$IQR = Q_3 - Q_1$$

Figure 1-8 shows the median and the interquartile range for two sets of numbers. One set has an odd number of elements, and the other set has an even number of elements.

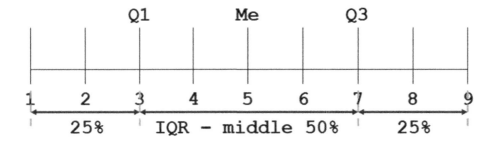

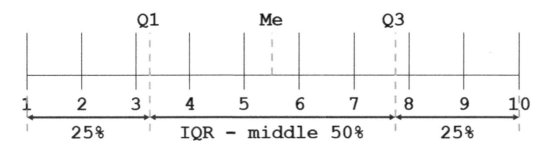

Figure 1-8. *Median and IQR*

Listing 1-7 shows the calculation of the IQR for the same two variables with the help of the PERCENTILE_CONT() window analytic function.

Listing 1-7. Calculating the IQR

```
-- IQR
SELECT DISTINCT
 PERCENTILE_CONT(0.75) WITHIN GROUP
  (ORDER BY 1.0*weightkg) OVER () -
 PERCENTILE_CONT(0.25) WITHIN GROUP
  (ORDER BY 1.0*weightkg) OVER () AS weightkgIQR,
 PERCENTILE_CONT(0.75) WITHIN GROUP
  (ORDER BY 1.0*hp) OVER () -
 PERCENTILE_CONT(0.25) WITHIN GROUP
  (ORDER BY 1.0*hp) OVER () AS hpIQR
FROM dbo.mtcars;
```

And here are the results.

```
weightkgIQR            hpIQR
---------------------- ----------------------
466.6325               83.5
```

The next possibility is to calculate the difference between every single value and the mean. This difference is also called the *deviation*. To get a single value, you can summarize the deviations. However, if some deviations are positive and some negative, the sum would be zero. Instead, you summarize squared deviations. To get a value that is closer to a single case value, you could divide the sum of squared deviations by the number of cases. However, this is not completely correct for a sample.

In the calculation, not every piece of information or every value varies. Imagine that you have a discrete variable with two possible states. If the first state frequency is 70%, then the second state frequency is not free; it is fixed. It can only be 30%.

With three states, one state is not free. If you know the frequencies for any two states, you can calculate the frequency of the third state. When you expand this to n states, you can conclude that only $n - 1$ states are free to vary. The sample with n cases has $n - 1$ *degrees of freedom*. Instead of dividing the sum of squared deviations by the number of cases, you divide it with the number of degrees of freedom, and you get the formula for the *variance*.

$$Var = \frac{1}{n-1} * \sum_{i=1}^{n}(v_i - \mu)^2$$

If you have data for the whole population, you can use the number of cases instead of the number of degrees of freedom. This way, you get the formula for *population variance*.

$$VarP = \frac{1}{n} * \sum_{i=1}^{n}(v_i - \mu)^2$$

If the number of cases is big, then $1/n$ is very close to $1/(n-1)$, and you can use any formula you wish for the sample or the population. But the value you get can still be very high because the formula uses squared deviations. You can calculate the squared root of the variance, and you get the formula for the standard deviation.

$$\sigma = \sqrt{Var}$$

Of course, you can calculate the standard deviation for the sample and the standard deviation for the population. In T-SQL, there are four functions to calculate the variance and the standard deviation for samples and populations, namely VAR(), VARP(), STDEV(), and STDEVP().

Standard deviation is also called the *second population moment* because every value comes in the formula on the second degree. It is hard to compare the standard deviations of two variables with different scales. For example, you might measure age in years and yearly income in thousands of dollars. Of course, the standard deviation for the income would be much higher than the standard deviation for the age. You can divide the standard deviation with the mean to get a relative measure called the *coefficient of variation*, as the following formula shows.

$$CV = \frac{\sigma}{\mu}$$

Now we have all the formulas needed to calculate the measures. Listing 1-8 shows the calculation in T-SQL.

Listing 1-8. Calculating variance, standard deviation, coefficient of variation

```
-- Variance, standard deviation, coefficient of variation
SELECT
 ROUND(VAR(weightkg), 2) AS weightVAR,
 ROUND(STDEV(weightkg), 2) AS weightSD,
 ROUND(VAR(hp), 2) AS hpVAR,
 ROUND(STDEV(hp), 2) AS hpSD,
 ROUND(STDEV(weightkg) / AVG(weightkg), 2) AS weightCV,
 ROUND(STDEV(hp) / AVG(hp), 2) AS hpCV
FROM dbo.mtcars;
```

Figure 1-9 shows the result of the query in Listing 1-8.

	weightVAR	weightSD	hpVAR	hpSD	weightCV	hpCV
1	196977.01	443.82	4700.87	68.56	0.3	0.47

Figure 1-9. *Variance, standard deviation, and coefficient of variation*

The hp variable has a bigger relative spread or a bigger coefficient of variation than the weightkg variable.

By comparing the median and the mean, you can conclude that there is a tail on one side. By comparing the IQR and the standard deviation, you can conclude that the tails might be on both sides. If the standard deviation is greater than the IQR, the distribution is tailed. There are important tails on both sides. If the skewness is positive for the same variable, the tail on the right side is even more important than the tail on the left side. For example, Listing 1-9 calculates the IQR and the standard deviation (SD) for the hp variable.

Listing 1-9. Query to Calculate the IQR and SD for the hp Variable

```
-- IQR and standard deviation for hp
SELECT DISTINCT
 PERCENTILE_CONT(0.75) WITHIN GROUP
  (ORDER BY 1.0*hp) OVER () -
 PERCENTILE_CONT(0.25) WITHIN GROUP
```

```
 (ORDER BY 1.0*hp) OVER () AS hpIQR,
 STDEV(hp) OVER() AS hpSD
FROM dbo.mtcars;
```

The results are shown in Figure 1-10.

	hpIQR	hpSD
1	83.5	68.5628684893206

Figure 1-10. *IQR and SD for the hp variable*

If you remember, the hp variable has an average greater than the median. Therefore, this variable has some tail on the right side. However, the standard deviation is lower than the IQR. Therefore, the tails on both sides are not really important. Can we get relative measures for the tail on one side and the overall "tailedness" of a distribution to compare these parameters of a distribution between two or more variables? I show these measures in the next section.

Skewness and Kurtosis

Skewness is a relative measure that tells you the importance of a tail on one side of a distribution curve. The following is the formula for skewness.

$$Skew = \frac{n}{(n-1)*(n-2)} * \sum_{i=1}^{n} \left(\frac{v_i - \mu}{\sigma} \right)^3$$

Skewness is also called the *third population moment* because the values come into the equation on the third degree. Because the deviations from the mean are already divided by the standard deviation, the skewness is a relative measure. You can compare skewness between two or more variables. A higher absolute value means a more skewed distribution. Positive skewness means a tail on the right side of the distribution. Negative skewness means a tail on the left side of the distribution.

The fourth population moment is *kurtosis*. You can imagine why the kurtosis is called the *fourth population moment*. Every value comes into the equation for kurtosis on the fourth degree. The following is the formula for kurtosis.

$$Kurt = \frac{n*(n+1)}{(n-1)*(n-2)*(n-3)} * \sum_{i=1}^{n} \left(\frac{v_i - \mu}{\sigma} \right)^4 - \frac{3*(n-1)^2}{(n-2)*(n-3)}$$

A positive kurtosis means that the tails are important. A negative means that the tails are not important. Like skewness, kurtosis is a relative measure.

When the formulas are clear, let's calculate these two higher population moments. Listing 1-10 performs this calculation for the hp variable.

Listing 1-10. Calculating Skewness and Kurtosis for a Variable

```
-- Skewness and kurtosis
WITH acs AS
(
SELECT hp,
 AVG(hp) OVER() AS a,
 COUNT(*) OVER() AS c,
 STDEV(hp) OVER() AS s
FROM dbo.mtcars
)
SELECT SUM(POWER((hp - a), 3) / POWER(s, 3)) / MIN(c)
 AS skewness,
 (SUM(POWER((hp - a), 4) / POWER(s, 4)) / MIN(c) -
3.0 * (MIN(c)-1) * (MIN(c)-1) / (MIN(c)-2) / (MIN(c)-3))
 AS kurtosis
FROM acs;
```

Figure 1-11 shows the results of the calculation. The variable has some tail on the right side because the skewness is positive. However, the tails are not important because the kurtosis is negative.

	skewness	kurtosis
1	0.755166486071029	-0.419639578319917

Figure 1-11. *Skewness and kurtosis of the hp variable*

I defined all the measures for continuous variables in this section. Let's put the most important ones together and calculate them in a single query. Listing 1-11 shows the query that calculates the count, the mean, the median, the standard deviation, the interquartile range, the coefficient of variation, the skewness, and the kurtosis for the hp variable.

Listing 1-11. Calculating All Important Measures for the hp Variable

```
-- All first population moments for hp
WITH acs AS
(
SELECT hp,
 AVG(hp) OVER() AS a,
 COUNT(*) OVER() AS c,
 STDEV(hp) OVER() AS s,
 PERCENTILE_CONT(0.5) WITHIN GROUP (ORDER BY hp) OVER () AS m,
 PERCENTILE_CONT(0.75) WITHIN GROUP (ORDER BY hp) OVER () -
 PERCENTILE_CONT(0.25) WITHIN GROUP (ORDER BY hp) OVER () AS i
FROM dbo.mtcars
)
SELECT MIN(c) AS hpCount,
 MIN(a) AS hpMean,
 MIN(m) AS hpMedian,
 ROUND(MIN(s), 2) AS hpStDev,
 MIN(i) AS hpIQR,
 ROUND(MIN(s) / MIN(a), 2) AS hpCV,
 ROUND(SUM(POWER((hp - a), 3) / POWER(s, 3))
           / MIN(c), 2) AS hpSkew,
 ROUND((SUM(POWER((hp - a), 4) / POWER(s, 4)) / MIN(c) -
           3.0 * (MIN(c)-1) * (MIN(c)-1)
           / (MIN(c)-2) / (MIN(c)-3)), 2) AS hpKurt
FROM acs;
GO
```

Figure 1-12 shows the results.

	hpCount	hpMean	hpMedian	hpStDev	hpIQR	hpCV	hpSkew	hpKurt
1	32	146	123	68.56	83.5	0.47	0.76	-0.42

Figure 1-12. *All first population moments for hp*

The query is written so that it is very simple to calculate all the aggregated values in groups. Listing 1-12 shows the query that calculates the same values for the hp variable, this time grouped by the engine variable.

Listing 1-12. All First Population Moments for hp Grouped by Engine

```
-- All first population moments for hp grouped by engine
WITH acs AS
(
SELECT engine,
 hp,
 AVG(hp) OVER (PARTITION BY engine) AS a,
 COUNT(*) OVER (PARTITION BY engine) AS c,
 STDEV(hp) OVER (PARTITION BY engine) AS s,
 PERCENTILE_CONT(0.5) WITHIN GROUP (ORDER BY hp)
  OVER (PARTITION BY engine) AS m,
 PERCENTILE_CONT(0.75) WITHIN GROUP (ORDER BY hp)
  OVER (PARTITION BY engine)-
 PERCENTILE_CONT(0.25) WITHIN GROUP (ORDER BY hp)
  OVER (PARTITION BY engine) AS i
FROM dbo.mtcars
)
SELECT engine,
 MIN(c) AS hpCount,
 AVG(hp) AS hpMean,
 MIN(m) AS hpMedian,
 ROUND(MIN(s), 2) AS hpStDev,
 MIN(i) AS hpIQR,
 ROUND(MIN(s) / MIN(a), 2) AS hpCV,
```

```
ROUND(SUM(POWER((hp - a), 3) / POWER(s, 3))
         / MIN(c), 2) AS hpSkew,
ROUND((SUM(POWER((hp - a), 4) / POWER(s, 4)) / MIN(c) -
         3.0 * (MIN(c)-1) * (MIN(c)-1)
         / (MIN(c)-2) / (MIN(c)-3)), 2) AS hpKurt
FROM acs
GROUP BY engine
ORDER BY engine;
GO
```

Figure 1-13 shows the results.

	engine	hpCount	hpMean	hpMedian	hpStDev	hpIQR	hpCV	hpSkew	hpKurt
1	Straight	14	91	96	24.42	43.75	0.27	-0.2	-2.46
2	V-shape	18	189	180	60.28	70	0.32	0.49	-0.74

Figure 1-13. *All first population moments for hp grouped by engine*

This was the last query in this section and this chapter, which is accumulating the knowledge needed to calculate the most important descriptive statistics values for continuous variables.

Conclusion

In this chapter, you learned how to calculate the summary values that quantitatively describe the variables from a dataset. You learned about the descriptive statistics calculations in T-SQL. For discrete variables, you calculate the frequencies. You should be very careful if the variables have an intrinsic order if they are ordinals. If ordinals are strings, you need to sort them properly. For ordinals, running totals make a lot of sense.

For continuous variables, you calculate many measures. You want to calculate more than a single measure for the center. In addition to the mean, you should always calculate the median. For the spread, calculate the standard population and the interquartile range. To compare multiple variables, relative measures are better than absolute ones. The coefficient of variability is better for comparisons than the standard deviation. Skewness and kurtosis are useful to get an impression about the tails on both sides of a distribution.

Before finishing this chapter, look at Figure 1-14.

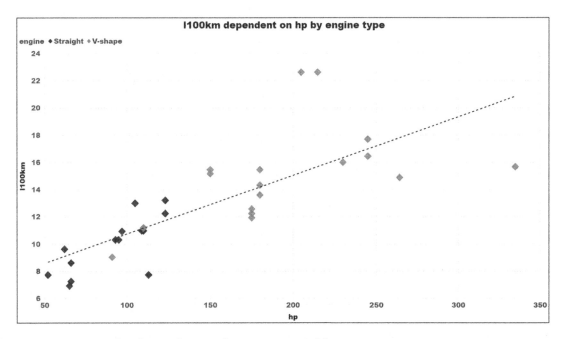

Figure 1-14. *The dependencies between variables*

Figure 1-14 is a scatterplot created with Power BI. It shows how the l100km variable (fuel consumption in liters per 100 kilometers) depends on the hp variable (horsepower) and the engine variable (engine type). The linear regression line for the association between the two continuous variables is shown too. Apparently, there are associations between these variables. In the next chapter, I explain the associations between pairs of variables and how to calculate them with T-SQL.

CHAPTER 2

Associations Between Pairs of Variables

After successfully analyzing distributions of single variables, let's move on to finding *associations* between pairs of variables. There are three possibilities.

- Both variables are continuous.

- Both variables are discrete.

- One variable is discrete, and the other one is continuous.

If two variables are associated, a change of the values in one variable means that the values of the other variable in the pair change at least partially synchronously and not completely randomly. You want to measure the *strength* of the association. In addition, you want to see whether the strength measured is *significant*. Therefore, you must introduce measures of significance.

Let's start with an example of two variables with only a few distinct numeric values. Table 2-1 shows the distribution of these two variables.

Table 2-1. *A Sample Distribution of Two Variables*

Value X_i	Probability $P(X_i)$	Value Y_i	Probability $P(Y_i)$
0	0.14	1	0.25
1	0.39	2	0.50
2	0.36	3	0.25
3	0.11		

© Dejan Sarka 2021
D. Sarka, *Advanced Analytics with Transact-SQL*, https://doi.org/10.1007/978-1-4842-7173-5_2

If the variables are truly independent, the distribution of Y should be the same over each X value and vice versa. The expected probability of every possible combination of both variables' values can be calculated as the product of separate probabilities of every value, as the following formula shows.

$$P(X_i, Y_i) = P(X_i) * P(Y_i)$$

You can calculate a table of all possible combinations and the expected probability. This is shown in Table 2-2.

Table 2-2. *Expected Probabilities If the Variables Were Independent*

X \ Y	1	2	3	P(X)
0	0.035	0.070	0.035	0.140
1	0.098	0.195	0.098	0.390
2	0.090	0.180	0.090	0.360
3	0.028	0.055	0.028	0.110
P(Y)	0.250	0.500	0.250	1 \ 1

But the observed probabilities could be quite different. Table 2-3 shows some fictious observed probabilities.

Table 2-3. *Observed Probabilities*

X \ Y	1	2	3	P(X)
0	0.140	0.000	0.000	0.140
1	0.000	0.260	0.130	0.390
2	0.000	0.240	0.120	0.360
3	0.110	0.000	0.000	0.110
P(Y)	0.250	0.500	0.250	1 \ 1

The two variables in this example are not completely independent.

In statistics, the starting point of an analysis is the *null hypothesis*. This is the default position that there is no association between two variables and that there is no difference in the characteristics of one variable considering the values of the other variable. Then you measure those differences and check how far away they are from the values when the two variables are mutually independent. You express the probability that the expected results for independent pairs of variables are at least as extreme as the results you observed and measured, considering that the null hypothesis is correct. This probability is the *p-value*. When the p-value is small enough—typically when it is lower than 0.05, you reject the null hypothesis and accept the possibility that the pair of variables is associated.

With the p-value, you calculate the area under the distribution function from the observed value toward the right or the left end of the curve. Figure 2-1 shows the area under the curve for standard normal distribution under the right tail when the z-value is greater than 1 and under both tails when the z-value is at least two standard deviations away from the mean.

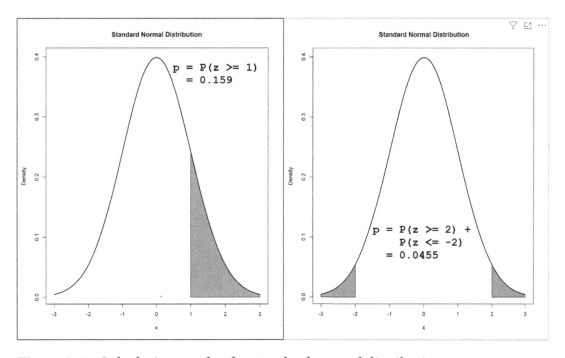

Figure 2-1. *Calculating p-value for standard normal distribution*

For the record, I created Figure 2-1 in Power BI by using the R script visual. Although this book is about T-SQL, you might be interested in this part of the R code as well, so it is shown here.

```
# Left visual
# Standard normal distribution
curve(dnorm(x), from = -3, to = 3,
      main = 'Standard Normal Distribution',
      ylab = 'Density',
      lwd = 2)
# Std norm dist density values
x <- seq(1, 3, by = 0.1)
y <- dnorm(x)
# Shade an area
polygon(c(x, rev(x)), c(y, rep(0, length(y))),
        col = adjustcolor("blue", alpha=0.3))

# Right visual
# Standard normal distribution
curve(dnorm(x), from = -3, to = 3,
      main = 'Standard Normal Distribution',
      ylab = 'Density',
      lwd = 2)
# Std norm dist density values
x <- seq(2, 3, by = 0.1)
y <- dnorm(x)
# Shade two areas
polygon(c(x, rev(x)), c(y, rep(0, length(y))),
        col = adjustcolor("blue", alpha=0.3))
x <- seq(-3, -2, by = 0.1)
y <- dnorm(x)
polygon(c(x, rev(x)), c(y, rep(0, length(y))),
        col = adjustcolor("blue", alpha=0.3))
```

You calculate the p-value with the definite integral. For the most important distributions in statistics, you can find precalculated tables. You can find those tables in practically any book about statistics and on many websites. There are even online calculators. However, you can calculate the definite integration with T-SQL, and you learn how to do it by the end of this chapter.

Associations Between Continuous Variables

To measure the strength of the relation between two variables, I use *covariance* as an absolute measurement and *correlation* as a relative one. I introduce the *coefficient of determination*, which is simply a squared correlation coefficient.

Let's start measuring dependency with covariance and using the real variables from the mtcars dataset.

Covariance

Figure 2-2 shows associations between two pairs of continuous variables in two *scatter charts*. The left chart shows the association between fuel consumption in liters per 100 kilometers dependent on horsepower. The right chart shows the miles per gallon you can drive dependent on horsepower. There is a positive association in the left chart and a negative association in the right chart. Positive association means that when the values of the independent variable rise, the values of the dependent variable go up. Negative association means that when the values of the independent variable rise, the values of the dependent variable go down.

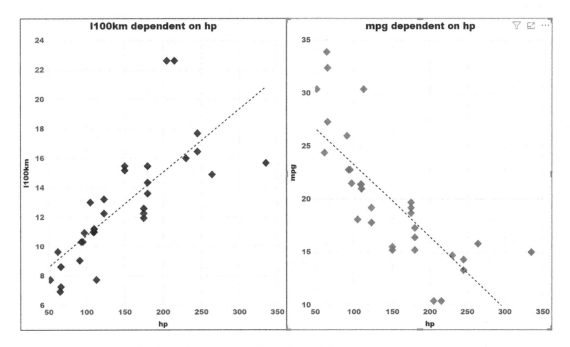

Figure 2-2. *Associations between pairs of variables*

The two dependent variables represent the same thing—fuel consumption, with two different measures. When I imported the mtcars dataset from R to SQL Server, I created a derived l100km variable from the source mpg variable. You could expect similar absolute values for the measures of the association but probably values with a different sign.

Let's start by introducing covariance with the formula for the variance, which was explained in Chapter 1.

$$Var = \frac{1}{n-1} * \sum_{i=1}^{n}\left(X_i - \mu(X)\right)^2$$

The formula summarizes the squares of the deviations of each value from the mean. Let's expand this formula to do manual multiplication for the square.

$$Var = \frac{1}{n-1} * \sum_{i=1}^{n}\left(X_i - \mu(X)\right)*\left(X_i - \mu(X)\right)$$

When dealing with a pair of variables, you replace one instance of the deviations from the mean of the first variable with the second variable. You get the formula for the covariance.

$$CoVar = \frac{1}{n-1} * \sum_{i=1}^{n}\left(X_i - \mu(X)\right)*\left(Y_i - \mu(Y)\right)$$

This is the formula for the sample. To get the formula for the population, you only need to replace the degrees of freedom $(n-1)$ with the number of cases (n).

$$CoVarP = \frac{1}{n} * \sum_{i=1}^{n}\left(X_i - \mu(X)\right)*\left(Y_i - \mu(Y)\right)$$

With this formula, it is easy to calculate the covariance for both pairs of variables introduced in Figure 2-2. Listing 2-1 shows this calculation. It uses the AVG() window aggregate function to add the means of all three variables to each row beside the original values in a common table expression. It then calculates the covariance for the sample in the outer query, using the COUNT(*) - 1 expression to calculate the number of degrees of freedom.

Listing 2-1. Calculating the Covariance

```
-- Covariance
-- l100km and mpg by hp
WITH CoVarCTE AS
(
SELECT l100km as val1,
 AVG(l100km) OVER() AS mean1,
 1.0 * hp AS val2,
 AVG(1.0 * hp) OVER() AS mean2,
 mpg AS val3,
 AVG(mpg) OVER() AS mean3
FROM dbo.mtcars
)
SELECT
 SUM((val1-mean1)*(val2-mean2)) / (COUNT(*)-1) AS Covar1,
 SUM((val3-mean3)*(val2-mean2)) / (COUNT(*)-1) AS Covar2
FROM CoVarCTE;
GO
```

Figure 2-3 shows the results.

	Covar1	Covar2
1	202.076754	-320.732056

Figure 2-3. *The two covariances*

You can see that the first covariance, the one between the l100km and hp variables, is positive, whereas the second covariance, between the mpg and hp variables, is negative. However, the second absolute value is much bigger than the first one. This is due to bigger absolute values in the mpg variable than in the l100km variable. It is hard to see which association is stronger. Since one dependent variable is derived from the other dependent variable, the strength of the association should be very similar. You need a relative measure for the strength of the association.

37

Correlation

When I wanted to calculate the relative measure for the spread, the coefficient of variation, I divided the standard deviation of a variable with the mean of the same variable. I can do something very similar to get a relative measure for the strength of the association between pairs of continuous variables. If I divide the covariance with the product of the standard deviations of both variables, I get the *correlation coefficient*, as the following formula shows.

$$\rho(X,Y) = \frac{\mathrm{CoVar}(X,Y)}{\sigma(X) * \sigma(Y)}$$

The correlation coefficient measures the amount of linear association between two continuous variables. The values are limited to the interval between –1 and 1, with borders included. A –1 value means perfect negative correlation, value 0 means no correlation, and value 1 means perfect positive correlation.

Before calculating the correlation coefficient, I want to introduce another useful measure. The *coefficient of determination* is simply the correlation coefficient squared. Usually, it is denoted with R^2.

$$R^2 = \rho(X,Y)^2$$

The coefficient of determination measures the proportion of the variance in the dependent variable that is predictable from or can be explained by the independent variable. This coefficient can take a value from the range between 0 and 1, borders included. Listing 2-2 shows the calculation of covariance, the correlation coefficient, and the coefficient of determination.

Listing 2-2. Calculating Covariance, Correlation, and Coefficient of Determination

```
-- l100km and mpg by hp
WITH CoVarCTE AS
(
SELECT l100km as val1,
 AVG(l100km) OVER () AS mean1,
 1.0 * hp AS val2,
 AVG(1.0 * hp) OVER() AS mean2,
```

```
 mpg AS val3,
 AVG(mpg) OVER() AS mean3
FROM dbo.mtcars
)
SELECT N'l100km by hp' AS Variables,
 SUM((val1-mean1)*(val2-mean2)) / (COUNT(*)-1) AS Covar,
 (SUM((val1-mean1)*(val2-mean2)) / (COUNT(*)-1)) /
 (STDEV(val1) * STDEV(val2)) AS Correl,
 SQUARE((SUM((val1-mean1)*(val2-mean2)) / (COUNT(*)-1)) /
 (STDEV(val1) * STDEV(val2))) AS CD
FROM CoVarCTE
UNION
SELECT N'mpg by hp' AS Variables,
 SUM((val3-mean3)*(val2-mean2)) / (COUNT(*)-1) AS Covar,
 (SUM((val3-mean3)*(val2-mean2)) / (COUNT(*)-1)) /
 (STDEV(val3) * STDEV(val2)) AS Correl,
 SQUARE((SUM((val3-mean3)*(val2-mean2)) / (COUNT(*)-1)) /
 (STDEV(val3) * STDEV(val2))) AS CD
FROM CoVarCTE;
```

The outer query makes a union of two result sets to get the coefficients for each pair of variables in its row. Figure 2-4 shows the result.

	Variables	Covar	Correl	CD
1	l100km by hp	202.076754	0.762911833930414	0.582034466351068
2	mpg by hp	-320.732056	-0.776168370733688	0.602437339727388

Figure 2-4. *Covariance, correlation, coefficient of determination*

The result shows that the correlation between the two pairs is very similar. Differences are due to a small sample. With a greater sample, you can also use the formulas for the population. Listing 2-3 shows the calculation for the population.

Listing 2-3. Calculating Covariance, Correlation, and Coefficient of
Determination for Population

```
-- Using formula for the population
-- l100km and mpg by hp
WITH CoVarCTE AS
(
SELECT
 l100km as val1,
 AVG(l100km) OVER () AS mean1,
 mpg as val3,
 AVG(mpg) OVER () AS mean3,
 1.0 * hp AS val2,
 AVG(1.0 * hp) OVER() AS mean2
FROM dbo.mtcars
)
SELECT N'l100km by hp' AS Variables,
 SUM((val1-mean1)*(val2-mean2)) / COUNT(*) AS Covar,
 (SUM((val1-mean1)*(val2-mean2)) / COUNT(*)) /
 (STDEVP(val1) * STDEVP(val2)) AS Correl,
 SQUARE((SUM((val1-mean1)*(val2-mean2)) / COUNT(*)) /
 (STDEVP(val1) * STDEVP(val2))) AS CD
FROM CoVarCTE
UNION
SELECT N'mpg by hp' AS Variables,
 SUM((val3-mean3)*(val2-mean2)) / COUNT(*) AS Covar,
 (SUM((val3-mean3)*(val2-mean2)) / COUNT(*)) /
 (STDEVP(val3) * STDEVP(val2)) AS Correl,
 SQUARE((SUM((val3-mean3)*(val2-mean2)) / COUNT(*)) /
 (STDEVP(val3) * STDEVP(val2))) AS CD
FROM CoVarCTE
ORDER BY Variables;
```

The results are shown in Figure 2-5.

	Variables	Covar	Correl	CD
1	l100km by hp	195.761855	0.762911832225415	0.582034463749539
2	mpg by hp	-310.709179	-0.776168370109175	0.602437338757933

Figure 2-5. *Covariance, correlation, coefficient of determination for population*

Even with only 32 cases, the results for the sample do not differ much from the results for the population.

Interpreting the Correlation

For the correlation coefficient, you cannot calculate the p-value. Therefore, you must use a rule of thumb for the interpretation of the significance of the coefficient. The interpretation depends on the content of the input variables. It is important to know what the input variables measure and how the values are gathered. If the values are gathered automatically, there are few errors. You can expect very high coefficients when there are associations between variables. When measuring social data, you can expect lower values for the correlation coefficients due to softer data. People can easily give you a different response to the same question if you ask it twice. Table 2-4 gives you a general idea of when the correlation coefficient and coefficient of determination are meaningful in different disciplines.

Table 2-4. *Meaningful Coefficients for a Discipline*

Discipline	ρ meaningful when	R^2 meaningful when
social	< -0.60 or > 0.60	> 0.35
biology	< -0.70 or > 0.70	> 0.50
chemistry	< -0.90 or > 0.90	> 0.80
physics	< -0.95 or > 0.95	> 0.90

The mean is likely the most frequently abused measure in statistics, and correlation is next to it. There are many possibilities for misinterpreting correlation. The correlation coefficient measures linear relationships only. There might be a perfect dependency between a pair of variables that is not linear, and you get a very low correlation coefficient.

Look at the following code.

```
-- Issues with non-linear relationships
CREATE TABLE #Nonlinear
 (x int,
  y AS SQUARE(x))
INSERT INTO #Nonlinear(x)
VALUES(-2), (-1), (0), (1), (2)
GO
DECLARE @mean1 decimal(10,6)
DECLARE @mean2 decimal(10,6)
SELECT @mean1=AVG(x*1.0),
       @mean2=AVG(y*1.0)
  FROM #Nonlinear
SELECT Correl=
       (SUM((x*1.0-@mean1)*(y*1.0-@mean2))
        /COUNT(*))
       /((STDEVP(x*1.0)*STDEVP(y*1.0)))
FROM #Nonlinear
DROP TABLE #Nonlinear;
GO
```

The code creates a simple temporary table with one independent variable and one dependent variable, which is a computed column. It computes the square of the first variable. The variables are associated. Yet, if you run the code, the calculation gives a correlation coefficient equal to zero.

Another big mistake is misinterpreting the correlation as the *causation*. Correlation does not say anything about causation. The correlation coefficient for a pair of variables is always the same, no matter which variable you declare as dependent or independent. Causation is just a matter of interpretation.

Finally, there are many *spurious correlations*. A spurious correlation can be a complete coincidence. An example of this is the US murder rate in the years between 2006 and 2011 dropping at the same rate as Microsoft Internet Explorer usage. Spurious correlation can also be caused by a third variable that was not measured but influences the measured variables used in calculations. For example, drownings rise when ice cream sales rise. But, people do not drown because they eat ice cream. Temperature is the hidden variable. With higher temperatures, people go swimming more frequently and drown more frequently. Of course, with higher temperatures, people also eat more ice cream. You should always be careful when you interpret correlation.

Associations Between Discrete Variables

When discovering the association between two discrete variables, you start with *contingency tables*. Contingency tables are pivot tables, also called *crosstabulations*, that show the actual and the expected frequencies if the pair of variables are mutually independent. Let's begin by examining associations graphically. Figure 2-6 shows a clustered column chart and a pivot table for examining the association between the power and the shape of an engine. The power is binned into three classes.

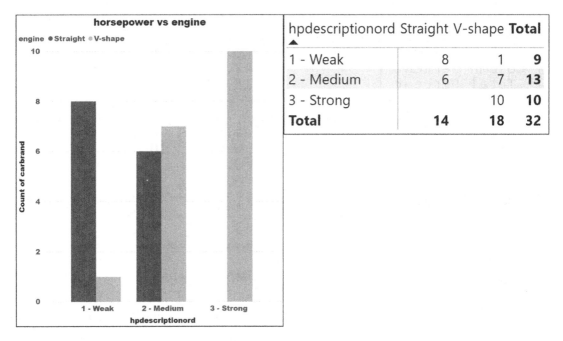

hpdescriptionord	Straight	V-shape	**Total**
1 - Weak	8	1	**9**
2 - Medium	6	7	**13**
3 - Strong		10	**10**
Total	**14**	**18**	**32**

Figure 2-6. *Graphical analysis of power vs. shape of an engine*

More powerful engines are usually V-shaped. A contingency table should show the same fact. But contingency tables have a bigger advantage than graphs: it is possible to calculate the strength of the association with a *chi-squared test* and the significance of the association with a p-value.

Contingency Tables

Contingency tables show observed frequencies with expected frequencies. They can show the case percentages and marginal percentages in a cell or the frequency distribution of every input variable separately. Figure 2-7 shows a typical contingency table.

```
|===================================================|
|                       engine                      |
|                  |hp      Straight   V-shape  Total|
| -----------------|--------------------------------
|            Count |Weak          8         1      9 |
| Expected Values  |            3.9       5.1         |
|     Row Percent  |          88.9%     11.1%   28.1%|
|  Column Percent  |          57.1%      5.6%         |
|   Total Percent  |          25.0%      3.1%         |
| -----------------|------------------------------- 
|            Count |Medium        6         7     13 |
| Expected Values  |            5.7       7.3         |
|     Row Percent  |          46.2%     53.8%   40.6%|
|  Column Percent  |          42.9%     38.9%         |
|   Total Percent  |          18.8%     21.9%         |
| -----------------|------------------------------- 
|            Count |Strong        0        10     10 |
| Expected Values  |            4.4       5.6         |
|     Row Percent  |           0.0%    100.0%   31.2%|
|  Column Percent  |           0.0%     55.6%         |
|   Total Percent  |           0.0%     31.2%         |
| -----------------|------------------------------- 
|                  |Total        14        18     32 |
|                  |           43.8%     56.2%        |
|===================================================|
```

Figure 2-7. *Contingency table for power and shape of engines*

Let's analyze the cell for the V-shaped engine with medium power. The count, or the observed frequency, is 7. For a pair of independent variables, 7.3 cases would be expected in this cell. The observed frequency (7) is 53.8% of the row count (13), 38.9% of the column count (18), and 21.9% of the total count (32). The marginal frequencies in the row totals tell you that you have 14 (42.8%) straight and 18 (56.2%) V-shaped engines, and in column totals that you have 9 (28.1%) weak, 12 (40.6%) medium, and 10 (31.2%) strong engines. Please note that in the cell for strong and straight engines, the observed frequency is zero.

It is not easy to reproduce a contingency table in T-SQL. I could try to mimic it with the PIVOT operator; however, this operator is very limited. It cannot accept a dynamic column list, and it cannot show more than one aggregate in a cell. In the contingency table in Figure 2-7, there are five different measures in each cell. I use the GROUP BY operator to get the desired analysis. Let's try the following query, which properly orders the power discretized variable.

```
-- Group by excludes empty rows
SELECT
 CASE hpdescription
      WHEN N'Weak' THEN N'1 - Weak'
      WHEN N'Medium' THEN N'2 - Medium'
      WHEN N'Strong' THEN N'3 - Strong'
     END AS X,
 engine AS Y,
 COUNT(*) AS obsXY
FROM dbo.mtcars
GROUP BY
 CASE hpdescription
      WHEN N'Weak' THEN N'1 - Weak'
      WHEN N'Medium' THEN N'2 - Medium'
      WHEN N'Strong' THEN N'3 - Strong'
     END
     ,engine
ORDER BY X, Y;
```

Here is the result.

```
X          Y          obsXY
---------- ---------- -----------
1 - Weak   Straight   8
1 - Weak   V-shape    1
2 - Medium Straight   6
2 - Medium V-shape    7
3 - Strong V-shape    10
```

Please note that the strong and straight engine group is missing because there is no case for it. However, I need this group to calculate the expected frequencies. I can generate all the groups I need if I cross join the distinct values of each independent variable.

```
-- Use a cross join to generate all rows
SELECT X, Y
FROM
(
SELECT DISTINCT
 CASE hpdescription
      WHEN N'Weak' THEN N'1 - Weak'
      WHEN N'Medium' THEN N'2 - Medium'
      WHEN N'Strong' THEN N'3 - Strong'
      END AS X
FROM dbo.mtcars) AS a
CROSS JOIN
(
SELECT DISTINCT
 engine AS Y
FROM dbo.mtcars) AS b;
```

I got the following result.

```
X          Y
---------- ----------
1 - Weak   Straight
2 - Medium Straight
3 - Strong Straight
1 - Weak   V-shape
2 - Medium V-shape
3 - Strong V-shape
```

Now let's put the two previous queries in common table expressions and then use the LEFT OUTER JOIN operator to join them and get all the groups. Listing 2-4 shows the query that aggregates the data and adds empty groups.

Listing 2-4. Returning Empty Rows When Grouping Data

```
-- Group by with all rows
WITH o1 AS
(
SELECT
 CASE hpdescription
        WHEN N'Weak' THEN N'1 - Weak'
        WHEN N'Medium' THEN N'2 - Medium'
        WHEN N'Strong' THEN N'3 - Strong'
       END AS X,
 engine AS Y,
 COUNT(*) AS obsXY
FROM dbo.mtcars
GROUP BY
 CASE hpdescription
        WHEN N'Weak' THEN N'1 - Weak'
        WHEN N'Medium' THEN N'2 - Medium'
        WHEN N'Strong' THEN N'3 - Strong'
       END
       ,engine
),
o2 AS
(
SELECT X, Y
FROM
(
SELECT DISTINCT
 CASE hpdescription
        WHEN N'Weak' THEN N'1 - Weak'
        WHEN N'Medium' THEN N'2 - Medium'
        WHEN N'Strong' THEN N'3 - Strong'
       END AS X
```

```
FROM dbo.mtcars) AS a
CROSS JOIN
(
SELECT DISTINCT
 engine AS Y
FROM dbo.mtcars) AS b
)
SELECT o2.X, o2.Y,
 ISNULL(o1.obsXY, 0) AS obsXY
FROM o2 LEFT OUTER JOIN o1
 ON o2.X = o1.X AND
    o2.Y = o1.Y
ORDER BY o2.X, o2.Y;
```

Figure 2-8 shows the result. Note that there is also a group for strong, straight engines.

	X	Y	obsXY
1	1 - Weak	Straight	8
2	1 - Weak	V-shape	1
3	2 - Medium	Straight	6
4	2 - Medium	V-shape	7
5	3 - Strong	Straight	0
6	3 - Strong	V-shape	10

Figure 2-8. *Grouping with empty groups*

Now let's add the expected frequencies and calculate the percentages. You can calculate expected frequencies, just like expected probabilities. Remember, the expected probability of every possible combination of the values of both variables can be calculated as the product of separate probabilities of every value of every input variable. If you use observed frequencies instead of probabilities, you get expected frequencies. Listing 2-5 shows the query that creates the contingency table.

Listing 2-5. Contingency Table with Expected Frequencies

```
-- Contingency table with chi-squared contribution
WITH o1 AS
(
SELECT
 CASE hpdescription
      WHEN N'Weak' THEN N'1 - Weak'
      WHEN N'Medium' THEN N'2 - Medium'
      WHEN N'Strong' THEN N'3 - Strong'
     END AS X,
 engine AS Y,
 COUNT(*) AS obsXY
FROM dbo.mtcars
GROUP BY
 CASE hpdescription
      WHEN N'Weak' THEN N'1 - Weak'
      WHEN N'Medium' THEN N'2 - Medium'
      WHEN N'Strong' THEN N'3 - Strong'
     END
     ,engine
),
o2 AS
(
SELECT X, Y
FROM
(
SELECT DISTINCT
 CASE hpdescription
      WHEN N'Weak' THEN N'1 - Weak'
      WHEN N'Medium' THEN N'2 - Medium'
      WHEN N'Strong' THEN N'3 - Strong'
     END AS X
```

```
FROM dbo.mtcars) AS a
CROSS JOIN
(
SELECT DISTINCT
 engine AS Y
FROM dbo.mtcars) AS b
),
obsXY_CTE AS
(
SELECT o2.X, o2.Y,
 ISNULL(o1.obsXY, 0) AS obsXY
FROM o2 LEFT OUTER JOIN o1
 ON o2.X = o1.X AND
    o2.Y = o1.Y
),
expXY_CTE AS
(
SELECT X, Y, obsXY
 ,SUM(obsXY) OVER (PARTITION BY X) AS obsX
 ,SUM(obsXY) OVER (PARTITION BY Y) AS obsY
 ,SUM(obsXY) OVER () AS obsTot
 ,CAST(ROUND(SUM(1.0 * obsXY) OVER (PARTITION BY X)
  * SUM(1.0 * obsXY) OVER (PARTITION BY Y)
  / SUM(1.0 * obsXY) OVER (), 2) AS NUMERIC(6,2)) AS expXY
FROM obsXY_CTE
)
SELECT X, Y,
 obsXY, expXY,
 ROUND(SQUARE(obsXY - expXY) / expXY, 2) AS chiSq,
 CAST(ROUND(100.0 * obsXY / obsX, 2) AS NUMERIC(6,2)) AS rowPct,
 CAST(ROUND(100.0 * obsXY / obsY, 2) AS NUMERIC(6,2)) AS colPct,
 CAST(ROUND(100.0 * obsXY / obsTot, 2) AS NUMERIC(6,2)) AS totPct
FROM expXY_CTE
ORDER BY X, Y;
GO
```

Figure 2-9 shows the result.

	X	Y	obsXY	expXY	chiSq	rowPct	colPct	totPct
1	1 - Weak	Straight	8	3.94	4.18	88.89	57.14	25.00
2	1 - Weak	V-shape	1	5.06	3.26	11.11	5.56	3.13
3	2 - Medium	Straight	6	5.69	0.02	46.15	42.86	18.75
4	2 - Medium	V-shape	7	7.31	0.01	53.85	38.89	21.88
5	3 - Strong	Straight	0	4.38	4.38	0.00	0.00	0.00
6	3 - Strong	V-shape	10	5.63	3.39	100.00	55.56	31.25

Figure 2-9. *Contingency table*

Please note the chiSq column. This column shows the contribution of each cell, or each combination of input values, to the chi-squared value. It's time to introduce this value.

Chi-Squared Test

The *chi-squared* value is the sum of the squared deviations of actual frequencies from expected frequencies, divided by the expected frequencies. The following is the formula.

$$\chi^2 = \sum \frac{(O-E)^2}{E}$$

The value by itself does not tell much without degrees of freedom. For a contingency table, the number of degrees of freedom is the product of the number of degrees of freedom for the variable on columns and the variable on rows. It is the number of rows minus one multiplied by the number of columns minus one of a contingency table, as the following formula shows.

$$DF = (C-1)*(R-1)$$

There is not a single chi-squared distribution; there is one for each number of degrees of freedom.

Figure 2-10 shows the probability density function graph for a couple of chi-squared probability density functions, for different number degrees of freedom on the left, and with a shaded region under the right tail for chi-squared greater than 14 for distribution with seven degrees of freedom.

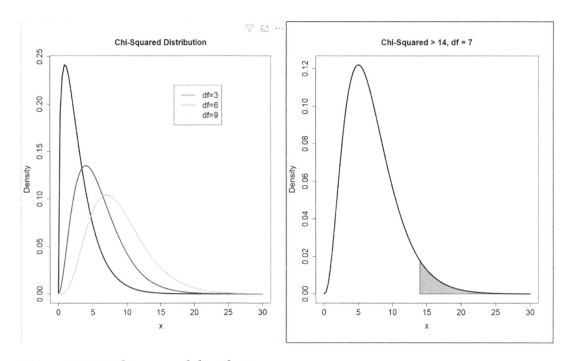

Figure 2-10. *Chi-squared distribution*

From the chi-squared distribution for a specific number of degrees of freedom, you can calculate the p-value with definite integration. Because these calculations are very frequent, it is easy to find tables with precalculated values or use online calculators. Table 2-5 shows precalculated values and the probability, or the area under the right tail of the probability density function, which is the p-value.

Table 2-5. *Chi-Squared Values and p-value*

DF	χ^2 value							
1	0.02	0.15	0.46	1.07	1.64	2.71	3.84	6.64
2	0.21	0.71	1.39	2.41	3.22	4.60	5.99	9.21
3	0.58	1.42	2.37	3.66	4.64	6.25	7.82	11.34
4	1.06	2.20	3.36	4.88	5.99	7.78	9.49	13.28
5	1.61	3.00	4.35	6.06	7.29	9.24	11.07	15.09
6	2.20	3.83	5.35	7.23	8.56	10.64	12.59	16.81
7	2.83	4.67	6.35	8.38	9.80	12.02	14.07	18.48
8	3.49	5.53	7.34	9.52	11.03	13.36	15.51	20.09
9	4.17	6.39	8.34	10.66	12.24	14.68	16.92	21.67
10	4.86	7.27	9.34	11.78	13.44	15.99	18.31	23.21
p-value	**0.90**	**0.70**	**0.50**	**0.30**	**0.20**	**0.10**	**0.05**	**0.01**
	Not significant					**Significant**		

Let's read an example from Table 2-5. If you have six degrees of freedom and the chi-squared value is greater than or equal to 12.59, the p-value is 0.05 or lower, meaning there is less than a 5% probability for such a big chi-squared value when the two variables analyzed are independent. You can safely reject the null hypothesis and say that the two variables are associated.

Let's calculate the chi-squared value and the degrees of freedom from the contingency table for the engine variables' power and shape. Listing 2-6 shows the query.

Listing 2-6. Calculating Chi-Squared and Degrees of Freedom

```
-- Chi Squared and DF
WITH o1 AS
(
SELECT
 CASE hpdescription
      WHEN N'Weak' THEN N'1 - Weak'
      WHEN N'Medium' THEN N'2 - Medium'
```

```
            WHEN N'Strong' THEN N'3 - Strong'
        END AS X,
 engine AS Y,
 COUNT(*) AS obsXY
FROM dbo.mtcars
GROUP BY
 CASE hpdescription
        WHEN N'Weak' THEN N'1 - Weak'
        WHEN N'Medium' THEN N'2 - Medium'
        WHEN N'Strong' THEN N'3 - Strong'
        END
        ,engine
),
o2 AS
(
SELECT X, Y
FROM
(
SELECT DISTINCT
 CASE hpdescription
        WHEN N'Weak' THEN N'1 - Weak'
        WHEN N'Medium' THEN N'2 - Medium'
        WHEN N'Strong' THEN N'3 - Strong'
        END AS X
FROM dbo.mtcars) AS a
CROSS JOIN
(
SELECT DISTINCT
 engine AS Y
FROM dbo.mtcars) AS b
),
obsXY_CTE AS
(
SELECT o2.X, o2.Y,
 ISNULL(o1.obsXY, 0) AS obsXY
```

```
FROM o2 LEFT OUTER JOIN o1
 ON o2.X = o1.X AND
    o2.Y = o1.Y
),
expXY_CTE AS
(
SELECT X, Y, obsXY
 ,SUM(obsXY) OVER (PARTITION BY X) AS obsX
 ,SUM(obsXY) OVER (PARTITION BY Y) AS obsY
 ,SUM(obsXY) OVER () AS obsTot
 ,ROUND(SUM(1.0 * obsXY) OVER (PARTITION BY X)
  * SUM(1.0 * obsXY) OVER (PARTITION BY Y)
  / SUM(1.0 * obsXY) OVER (), 2) AS expXY
FROM obsXY_CTE
)
SELECT SUM(ROUND(SQUARE(obsXY - expXY) / expXY, 2)) AS ChiSquared,
 (COUNT(DISTINCT X) - 1) * (COUNT(DISTINCT Y) - 1) AS DegreesOfFreedom
FROM expXY_CTE;
GO
```

The result of the query in Listing 2-6 is shown in Figure 2-11.

	ChiSquared	DegreesOfFreedom
1	15.24	2

Figure 2-11. *Chi-squared and degrees of freedom*

Table 2-5 shows that the p-value is lower than 0.01. Therefore, you can safely assume that the engine power and engine shape variables are highly associated.

Note   If you are interested in more information on chi-squared distribution, refer to the article at `https://en.wikipedia.org/wiki/Chi-square_distribution`.

Associations Between Discrete and Continuous Variables

One option remains for associations between two variables: one variable is discrete, and the other is continuous. You can analyze descriptive statistics population moments, for example mean or standard deviation, of the continuous variable in groups of the discrete variable. The null hypothesis, where you assume there is no association, means that there are no differences in population moments of the continuous variable in different classes of the discrete variable. However, many times it is easy to spot the differences.

Figure 2-12 shows two graphs analyzing the weight (in kilograms) of automatic and manual transmissions. You can see the differences between the two groups in both charts.

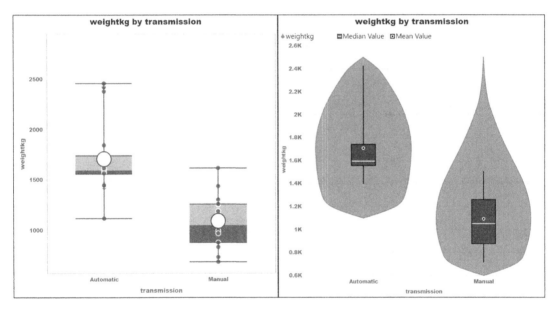

Figure 2-12. *Analyzing weight by transmission*

The left chart is called a *box plot* (or boxplot). The right chart is called a *violin plot*. The box plot shows the mean (the circle), the median (the line that splits the box), the IQR (the box's height), and the range of the weightkg variable in classes of the transmission variable. The violin plot shows the mean, the median, the IQR, and the distribution function shape of weight in classes of transmission. This distribution is shown twice—on the left and right sides of the boxes and the middle vertical line in a group—to make a shape that resembles a violin.

Testing Continuous Variable Moments over a Discrete Variable

In T-SQL, you can test any population moment of a continuous variable in the classes of a discrete one. In Chapter 1, I developed a query that does these tests. Let's now analyze weight by transmission class, as shown in Listing 2-7.

Listing 2-7. Analyzing weightkg Moments Transmission Classes

```
-- All first population moments for weightkg grouped by
-- transmission
WITH acs AS
(
SELECT transmission,
 weightkg,
 AVG(weightkg) OVER (PARTITION BY transmission) AS a,
 COUNT(*) OVER (PARTITION BY transmission) AS c,
 STDEV(weightkg) OVER (PARTITION BY transmission) AS s,
 PERCENTILE_CONT(0.5) WITHIN GROUP (ORDER BY weightkg)
  OVER (PARTITION BY transmission) AS m,
 PERCENTILE_CONT(0.75) WITHIN GROUP (ORDER BY weightkg)
  OVER (PARTITION BY transmission)-
 PERCENTILE_CONT(0.25) WITHIN GROUP (ORDER BY weightkg)
  OVER (PARTITION BY transmission) AS i
FROM dbo.mtcars
)
SELECT transmission,
 MIN(c) AS wCount,
 AVG(weightkg) AS wMean,
 MIN(m) AS wMedian,
 ROUND(MIN(s), 2) AS wStDev,
 MIN(i) AS wIQR,
 ROUND(MIN(s) / MIN(a), 2) AS wCV,
 ROUND(SUM(POWER((weightkg - a), 3) / POWER(s, 3)) / MIN(c), 2)
       AS wSkew,
```

```
ROUND((SUM(POWER((weightkg - a), 4) / POWER(s, 4)) / MIN(c) -
      3.0 * (MIN(c)-1) * (MIN(c)-1) / (MIN(c)-2)
      / (MIN(c)-3)), 2) AS wKurt
FROM acs
GROUP BY transmission
ORDER BY transmission;
```

Figure 2-13 shows the results.

	transmission	wCount	wMean	wMedian	wStDev	wIQR	wCV	wSkew	wKurt
1	Automatic	19	1709.540526	1596.64	352.62	183.7	0.21	0.98	-0.43
2	Manual	13	1093.610000	1052.33	279.86	383.29	0.26	0.21	-2.1

Figure 2-13. *Analyzing a continuous variable in classes of a discrete one*

You can see the differences in the mean and the median. Cars with automatic transmissions tend to weigh more. However, the IQR is much higher in a manual transmission than in an automatic transmission, meaning that there is an outlier—a car with a manual transmission that is not very light. If you check the data, this car is a Maserati Bora.

Analysis of Variance

Now you know that there is a difference in weight moments, such as the mean or different transmissions. However, the question is whether this difference is due to different transmissions or if it comes from the variability within groups. For example, a single extremely heavy car with an automatic transmission would substantially increase this transmission group; yet, this would not mean that cars with an automatic transmission weigh more than cars with a manual transmission. You need to compare the variance in the mean between groups with the variance within groups. You can calculate the variance between groups with the following formula.

$$MS_A = \frac{SS_A}{DF_A},$$

$$where \; SS_A = \sum_{i=1}^{a} n_i * \left(\mu_i - \mu \right)^2,$$

$$and \; DF_A = \left(a - 1 \right)$$

The following describes the elements in the formula.

- MS_A is the variance of the mean between groups.

- SS_A is the sum of squared differences between the group means and the overall mean.

- a is the number of groups or the number of distinct values in the discrete variable.

- n_i is the number of rows in each group.

- μ is the mean of the continuous variable over the whole sample.

- μ_i is the mean of the continuous variable in the i-th group of the discrete variable.

- DF_A is the groups' degrees of freedom, which is the number of groups minus one.

The variance within groups is defined as the normal variance, which is the sum of the squared deviations for each value from the group mean, and then summarizes the groups overall. The number of degrees of freedom is also the sum of the degrees of freedom in each group. The following shows the formula.

$$MS_E = \frac{SS_E}{DF_E},$$

$$where \; SS_E = \sum_{i=1}^{a} \sum_{j=1}^{ni} \left(v_{ij} - \mu_i \right)^2,$$

$$and \; DF_E = \sum_{i=1}^{a} \left(n_i - 1 \right)$$

The following describes the elements in the formula.

- MS_E is the variance of the mean within groups.

- SS_E is the sum of the squared differences between each value and the group mean or deviations within groups.

- a is the number of groups or the discrete variable's number of distinct values.

- n_i is the number of rows in each group.

- DF_E is the number of degrees of freedom within groups.

- v_{ij} is the value of the continuous variable in row j of group i.

- μ_i is the mean of the continuous variable in a group.

The measure for the strength of the association is called the *F-ratio*. As the following formula shows, this is simply the variance between groups divided by the variance within groups.

$$F = \frac{MS_A}{MS_E}$$

There are different F-distributions for the number of pairs of degrees of freedom between and within groups. Figure 2-14 shows examples of the F-distribution for different number degrees of freedom on the left. The figure on the right has a shaded region under the right tail for F-ratio greater than two. The discrete variabl has five disting values and thus four degrees of freedom and the continuous variable has ten degrees of freedom.

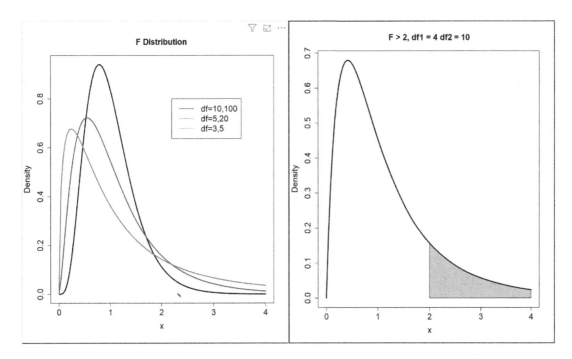

Figure 2-14. *F-ratio distributions*

Note   If you are interested in more information on F-distributions, refer to the article at https://en.wikipedia.org/wiki/F-distribution.

Now let's analyze the variance of the mean of the continuous variable in groups of the discrete variable, or the *ANOVA*. Listing 2-8 shows the query.

Listing 2-8. ANOVA

```
-- One-way ANOVA
WITH Anova_CTE AS
(
SELECT transmission, weightkg,
 COUNT(*) OVER (PARTITION BY transmission) AS gr_CasesCount,
 DENSE_RANK() OVER (ORDER BY transmission) AS gr_DenseRank,
 SQUARE(AVG(weightkg) OVER (PARTITION BY transmission) -
        AVG(weightkg) OVER ()) AS between_gr_SS,
 SQUARE(weightkg -
```

```
        AVG(weightkg) OVER (PARTITION BY transmission))
        AS within_gr_SS
FROM dbo.mtcars
)
SELECT N'Between groups' AS [Source of Variation],
 MAX(gr_DenseRank) - 1 AS df,
 ROUND(SUM(between_gr_SS), 0) AS [Sum Sq],
 ROUND(SUM(between_gr_SS) / (MAX(gr_DenseRank) - 1), 0)
  AS [Mean Sq],
 ROUND((SUM(between_gr_SS) / (MAX(gr_DenseRank) - 1)) /
 (SUM(within_gr_SS) / (COUNT(*) - MAX(gr_DenseRank))), 2)
  AS F
FROM Anova_CTE
UNION
SELECT N'Within groups' AS [Source of Variation],
 COUNT(*) - MAX(gr_DenseRank) AS Df,
 ROUND(SUM(within_gr_SS), 0) AS [Sum Sq],
 ROUND(SUM(within_gr_SS) / (COUNT(*) - MAX(gr_DenseRank)), 0)
  AS [Mean Sq],
 NULL AS F
FROM Anova_CTE;
```

After a long explanation, the query itself is not too complex. Note the DENSE_RANK() window ranking function. It calculates the degrees of freedom between and within groups. Figure 2-15 shows the results.

	Source of Variation	df	Sum Sq	Mean Sq	F
1	Between groups	1	2928265	2928265	27.64
2	Within groups	30	3178022	105934	NULL

Figure 2-15. *ANOVA results*

You can use either an online calculator to calculate the p-value for the F-value with the calculated degrees of freedom or find a table with precalculated F-values. However, because there is a pair of degrees of freedom, these tables are huge and impractical.

For practical purposes, I created a simple Visual C# console application called FDistribution.exe to calculate the p-value for the F-distribution. I do not show it here, but the code and the executable application are included in this book's companion content. You can call the console application in SQLCMD mode directly from SQL Server Management Studio, as the following code shows, assuming the executable is in the C:\temp folder.

```
!!C:\temp\FDistribution 27.64 1 30
GO
```

The result shows again that the p-value is lower than 0.01.

Definite Integration

There is another option for calculating the p-value. When you know the distribution of a variable and the distribution formula, you can store the values of the probability density function in a table together with the values of the variable and then calculate the definite integral for a specific interval between two values of the variable. No function in T-SQL would calculate the definite integral. But it is easy to use a *trapezoidal rile* for approximate calculation. The following is the formula.

$$\int_a^b f(x)dx \approx \frac{(b-a)}{2}(f(a)+f(b))$$

Figure 2-16 is a graphical explanation of the formula.

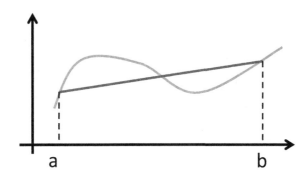

Figure 2-16. *Trapezoidal rule for definite integration*

You can split the area under the curve into multiple smaller areas or multiple smaller trapeziums. The narrower the trapeziums, the more accurate the approximate calculation of the definite integral. With the following formula, you can expand the trapezoidal rule on multiple areas with the same distance h between points on the X axis.

$$\int_a^b f(x)\,dx \approx \frac{h}{2}[f(x_1)+f(x_n)+2*(f(x_2)+f(x_3)+\ldots+f(x_{n-1}))]$$

This formula can be easily implemented in T-SQL. Let's first create a demo table for the standard normal distribution, where the mean is at 0, and the standard deviation is at 1. You can check the distribution function shape again in Figure 2-1.

Listing 2-9 shows the code that creates a temporary table and inserts the points with the 0.01 distance between points.

Listing 2-9. Table with Standard Normal Distribution Values

```
-- Standard normal distribution table
CREATE TABLE #t1
(z DECIMAL(3,2),
 y DECIMAL(10,9));
GO
-- Insert the data
SET NOCOUNT ON;
GO
DECLARE @z DECIMAL(3,2), @y DECIMAL(10,9);
SET @z=-4.00;
WHILE @z <= 4.00
  BEGIN
   SET @y=1.00/SQRT(2.00*PI())*EXP((-1.00/2.00)*SQUARE(@z));
   INSERT INTO #t1(z,y) VALUES(@z, @y);
   SET @z=@z+0.01;
END
GO
SET NOCOUNT OFF;
GO
-- Check the data
```

```
SELECT *
FROM #t1;
GO
```

Now let's calculate the area under the curve for the z-values between 0 and 1. Then, calculate the area under the right tail for z-values greater than or equal to 1.96. Listing 2-10 shows this code.

Listing 2-10. Using the Trapezoidal Rule for Definite Integration

```
-- Trapezoidal rule for definite integration
-- Pct of area between 0 and 1
WITH z0 AS
(
SELECT z, y,
  FIRST_VALUE(y) OVER(ORDER BY z) AS fy,
  LAST_VALUE(y)
   OVER(ORDER BY z
        ROWS BETWEEN UNBOUNDED PRECEDING
        AND UNBOUNDED FOLLOWING) AS ly
FROM #t1
WHERE z >= 0 AND z <= 1
)
SELECT 100.0 * ((0.01 / 2.0) *
 (SUM(2 * y) - MIN(fy) - MAX(ly))) AS pctdistribution
FROM z0;
-- Right tail after z >= 1.96
WITH z0 AS
(
SELECT z, y,
  FIRST_VALUE(y) OVER(ORDER BY z) AS fy,
  LAST_VALUE(y)
   OVER(ORDER BY z
        ROWS BETWEEN UNBOUNDED PRECEDING
        AND UNBOUNDED FOLLOWING) AS ly
```

```
FROM #t1
WHERE (z >= 0 AND z <= 1.96)
)
SELECT 50 - 100.0 * ((0.01 / 2.0) *
 (SUM(2 * y) - MIN(fy) - MAX(ly))) AS pctdistribution
FROM z0;
GO
```

Here are the results.

```
pctdistribution
---------------------------------------
34.134270

pctdistribution
---------------------------------------
2.499880
```

In the normal distribution, more than two-thirds of the values are less than one standard deviation away from the mean. And if you go approximately two standard deviations away from the mean in both directions (precisely 1.96 times the standard deviation away), about 5% of values are under both the left and right tails. This is where the "magic" of p-value 0.05 comes from. The value is typically the cutoff value for rejecting the null hypothesis.

Conclusion

The calculations and queries in this chapter became more complex. However, they revealed some interesting associations. This chapter used two variables for any single analysis and demonstrated a bivariate analysis. Later in this book, you see how to do multivariate analyses, where you must deal with more than two variables.

So far, you have played with neat, clean data and did not have to do much data preparation. However, in a real analytical project with business data, you typically spend more than half the time working on data preparation and data cleansing. This work is unfortunately not so interesting as the analyzing work is. Nevertheless, data preparation is many times more important than the analysis itself. In the next chapter, I explain many data preparation tasks.

PART II

Data Preparation and Quality

CHAPTER 3

Data Preparation

In any analytical or business intelligence (BI) project, data preparation is crucial. It might also be the longest part of a project—exhausting and sometimes tedious. However, the success of a project heavily depends on data preparation.

To appropriately prepare it for the project target, you need to know your data. Besides knowing the variables' distributions, you need to know the business meaning of your variables. It is always a good idea to have a subject expert available. Only with subject knowledge can create meaningful *derived variables*. Using carefully crafted derived variables can give you much better insights into the problem you are solving than using variables that come directly from a line of business (LOB) system. For example, basic height and weight variables do not reveal much about a person's health; however, the *obesity index* (height^2/weight) is very meaningful.

All the values available are now always available, so you may need to deal with missing values. There are situations in which there are too many missing values to do a valuable analysis.

When you get character data from your source system, such as names, descriptions, and addresses, these strings can be inconsistent or a denormalized, meaning that a single string represents multiple scalar values. You might need to extract the scalar values or make the strings consistent.

Grouping data might be part of the initial data overview, the data preparation, or the final analysis. Since this is an advanced book, I do not talk about the basics of the GROUP BY clause. Instead, I show you how to do efficient grouping over multiple variables with the GROUPING SETS clause.

Part of data preparation is the *data normalization* of continuous numeric variables. Data normalization means bringing variables to the same scale, so you have comparable values; for example, measuring yearly income in thousands of dollars or euros or measuring age in years. Having the same scale for both variables might improve some analyses.

69

© Dejan Sarka 2021
D. Sarka, *Advanced Analytics with Transact-SQL*, https://doi.org/10.1007/978-1-4842-7173-5_3

For some analyses, you use only numbers. If you want to use string variables, you need to *convert* them to numerical variables or numerics. For other analyses, you want to use only discrete variables. To use continuous variables, you need to *discretize* them.

In a real-life project, you might encounter many other data preparation tasks. This chapter overviews the most common tasks; it is not an exhaustive list of all possible data preparation tasks. Specialized languages like R and Python have many out-of-the-box data preparation functions or are available through libraries. T-SQL code might be more complex. However, when you deal with huge amounts of data, you want to use a mature database engine like SQL Server because this engine has been optimized to work with data for decades, and it excels in this area.

Dealing with Missing Values

Missing values come from the source systems. It is empty, nonexistent, or uncollected data. If there are not many missing values, and there is no *pattern* in the missing values, your dataset might be useful. The following are some options for how to deal with missing values.

- Filter the rows containing the missing data in any of the columns. This is a good solution if there are not too many missing values and there is no pattern in them.

- Ignore the column. This is less useful than filtering the rows because you typically have a lot of rows; however, the number of columns is still limited.

- Build separate models with rows and columns with missing data and without them. Some analyses might work well with missing values included.

- Modify the operational systems so that you can collect the missing values. This is an ideal solution. Usually, this solution cannot be achieved simply and reasonably. Modifying LOB systems might take years of time and enormous resources.

- Replace missing data with common (e.g., mean) values. This is a popular method for numerical variables. You can also use this method for discrete variables by taking the most frequent state.

You are not changing the four population moments often, but there is also a drawback. You are changing the measures for the associations between variables; hopefully, any missing variables are randomly spread. When you replace them with the mean, for example, two associated variables covariate less.

- Predict new values. This is a method similar to replacing missing values with common values, just more complex. You use multiple other variables to predict the value of the missing one.

In a SQL Server table, the NULL placeholder is used for a missing value. Before dealing with missing data, you need to understand how SQL Server treats NULLs in functions and other code elements.

NULLs in T-SQL Functions

Let's start by creating a simple table in the AdventureWorksDW2017 demo database with the following code.

```
USE AdventureWorksDW2017;
DROP TABLE IF EXISTS #t1;
CREATE TABLE #t1
(
 col1 INT NULL,
 col2 INT NULL,
 col3 INT NULL
);
GO
```

Next, let's populate this table. Listing 3-1 shows the code for inserting a few rows and checking the data.

Listing 3-1. Data with NULLs

```
INSERT INTO #t1 VALUES
(2, NULL, 6),
(8, 10, 12),
(NULL, 16, 18),
```

```
(20, 22, 24),
(26, NULL, NULL);
SELECT *
FROM #t1;
```

Figure 3-1 shows the result.

	col1	col2	col3
1	2	NULL	6
2	8	10	12
3	NULL	16	18
4	20	22	24
5	26	NULL	NULL

Figure 3-1. *A table with NULLs*

The ISNULL() T-SQL function accepts two parameters: the column and the replacement value. It returns the column value when the value is known and the replacement value when the column value is NULL. If all the inputs are NULL, NULL is returned.

The COALESCE() function accepts multiple parameters and returns the first non-NULL value. The aggregate functions simply skip the missing values. The only exception is the COUNT(*) function, which counts all rows, no matter the missing values. Listing 3-2 shows how these functions work.

Listing 3-2. T-SQL Functions on Data with Nulls

```
-- ISNULL and COALESCE
SELECT col1,
 ISNULL(col1, 0) AS c1NULL,
 col2, col3,
 COALESCE(col2, col3, 99) AS c2NULL
FROM #t1;
-- Aggregate functions
SELECT
 AVG(col2) AS c2AVG,
 SUM(col2) AS c2SUM,
```

```
COUNT(*) AS n,
 SUM(1.0*col2)/COUNT(*) AS col2SumByCount
FROM #t1;
GO
```

Figure 3-2 shows the results.

	col1	c1NULL	col2	col3	c2NULL
1	2	2	NULL	6	6
2	8	8	10	12	10
3	NULL	0	16	18	16
4	20	20	22	24	22
5	26	26	NULL	NULL	99

	c2AVG	c2SUM	n	col2SumByCount
1	16	48	5	9.600000

Figure 3-2. *Results of T-SQL functions on missing data*

In the third row of the upper result, you can see how the ISNULL() function returned 0 when col1 was equal to NULL. In rows 1 and 5, you see how the COALESCE() function works. The lower result shows that the mean value calculated by the AVG() function differs for the value calculated with the SUM() function and divided by COUNT(*). The AVG() and SUM() functions skip the rows where the input column is NULL.

Handling NULLs

I found a nice dataset in R to show you how to handle NULLs. The air quality dataset (www.rdocumentation.org/packages/datasets/versions/3.6.2/topics/airquality) is simple, yet it has NULLs in two variables. The dataset shows daily air quality measurements in New York, NY, from May to September 1973. The following describes the four measurement variables.

- Ozone: Mean ozone in parts per billion from 13:00 to 15:00 hours on Roosevelt Island

- SolarR: Solar radiation in Langleys in the frequency band of 4000–7700 angstroms from 08:00 to 12:00 hours in Central Park

- Wind: Average wind speed in miles per hour at 07:00 and 10:00 hours
  at LaGuardia Airport

- Temp: Maximum daily temperature in degrees Fahrenheit at La
  Guardia Airport

The other two variables represent the month and day. All variables are integers. Listing 3-3 shows the code that creates a table and loads the dataset. Since the table's purpose is to show NULL handling, I used only integer data types in all columns.

Listing 3-3. Loading the R Air Quality Dataset

```
-- Load the airquality dataset
DROP TABLE IF EXISTS dbo.airquality;
CREATE TABLE dbo.airquality
(
 Ozone int,
 SolarR int,
 Wind int,
 Temp int,
 MonthInt int,
 DayInt int
);
GO
INSERT INTO dbo.airquality
EXECUTE sys.sp_execute_external_script
 @language=N'R',
 @script = N'
data("airquality")
 ',
 @output_data_1_name = N'airquality';
GO
SELECT *
FROM dbo.airquality;
GO
```

In the last SELECT statement, the first two variables, Ozone and SolarR have NULLs. Figure 3-3 shows the partial result of the query.

	Ozone	SolarR	Wind	Temp	MonthInt	DayInt
1	41	190	7	67	5	1
2	36	118	8	72	5	2
3	12	149	12	74	5	3
4	18	313	11	62	5	4
5	NULL	NULL	14	56	5	5
6	28	NULL	14	66	5	6
7	23	299	8	65	5	7

Figure 3-3. *The air quality dataset*

The following code checks for the number of NULLs in the first two variables.

```
SELECT IIF(Ozone IS NULL, N'NULL', N'Not NULL') AS OzoneN,
 COUNT(*) AS cnt
FROM dbo.airquality
GROUP BY IIF(Ozone IS NULL, N'NULL', N'Not NULL');
SELECT IIF(SolarR IS NULL, N'NULL', N'Not NULL') AS SolarRN,
 COUNT(*) AS cnt
FROM dbo.airquality
GROUP BY IIF(SolarR IS NULL, N'NULL', N'Not NULL');
GO
```

There are 37 NULLs in the Ozone variable and 7 in the SolarR variable. Now let's check how many rows have NULLs in at least one variable. The following code adds an indicator variable and checks the total number of rows with NULLs.

```
-- Adding indicator variable for missing values
SELECT *,
 IIF(IIF(Ozone IS NULL, 1, 0) = 1 OR
     IIF(SolarR IS NULL, 1, 0) = 1, 1, 0)
 AS NullInRow
FROM dbo.airquality;
-- Number of rows with NULLs
```

```
SELECT
 IIF(IIF(Ozone IS NULL, 1, 0) = 1 OR
     IIF(SolarR IS NULL, 1, 0) = 1, 1, 0)
 AS NullInRow,
 COUNT(*) AS cnt
FROM dbo.airquality
GROUP BY
 IIF(IIF(Ozone IS NULL, 1, 0) = 1 OR
     IIF(SolarR IS NULL, 1, 0) = 1, 1, 0);
```

The second query tells us that there are NULLs in 42 rows out of 153 total rows. This is a huge number. Still, those rows might be filtered if there is no pattern in the NULLs. Listing 3-4 shows the code for this check.

Listing 3-4. Checking for Patterns with NULLs

```
-- Check the other variables in the classes of the NULL indicator
SELECT
 IIF(IIF(Ozone IS NULL, 1, 0) = 1 OR
     IIF(SolarR IS NULL, 1, 0) = 1, 1, 0)
 AS NullInRow,
 AVG(1.0 * Wind) AS WindA,
 AVG(1.0 * Wind) / STDEV(1.0 * Wind) AS WindCV,
 AVG(1.0 * Temp) AS TempA,
 AVG(1.0 * Temp) / STDEV(1.0 * Temp) AS TempCV
FROM dbo.airquality
GROUP BY
 IIF(IIF(Ozone IS NULL, 1, 0) = 1 OR
     IIF(SolarR IS NULL, 1, 0) = 1, 1, 0);
```

The code checks the mean and coefficient of variation for the Wind and Temp variables in the NULL indicator variable classes. Figure 3-4 shows that there are small differences in the results.

	NullInRow	WindA	WindCV	TempA	TempCV
1	0	9.477477	2.69370088307346	77.792792	8.16296371052822
2	1	9.476190	2.73978651150371	78.119047	8.30865103203644

Figure 3-4. *No big differences in classes of the NULL indicator*

I also checked the mean of the Temp and the Wind variables in the classes of the NULL indicator variable, as shown in Figure 3-5.

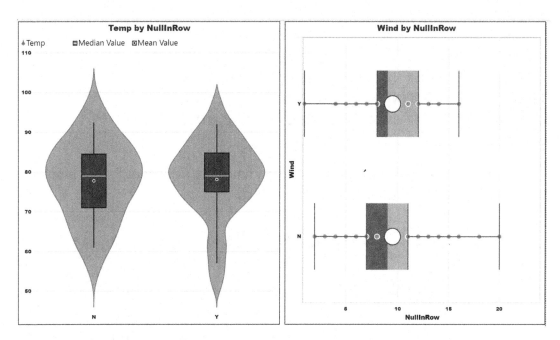

Figure 3-5. *Graphically checking for patterns in NULLs*

Filtering all rows with NULLs in any of the columns is also called *listwise* filtering. Listing 3-5 shows how to do it.

Listing 3-5. Listwise Filtering

```
-- Listwise filtering rows with NULLs
SELECT *
FROM dbo.airquality
WHERE
 IIF(IIF(Ozone IS NULL, 1, 0) = 1 OR
    IIF(SolarR IS NULL, 1, 0) = 1, 1, 0) = 0;
```

Replacing NULLs, or *imputing* values, would probably impact the Ozone variable because there are many NULLs in this variable. However, this would make sense for the SolarR variable. Listing 3-6 shows how you can impute the mean or the median value in place of NULLs.

Listing 3-6. Imputing Values

```
SELECT SolarR,
 AVG(SolarR) OVER() AS SolarRA,
 ISNULL(SolarR, AVG(SolarR) OVER()) AS SolarRIA,
 PERCENTILE_CONT(0.5)
  WITHIN GROUP (ORDER BY SolarR) OVER() AS SolarM,
 ISNULL(SolarR,  PERCENTILE_CONT(0.5)
  WITHIN GROUP (ORDER BY SolarR) OVER()) AS SolarIM
FROM dbo.airquality;
```

The ISNULL() function, together with window functions, was handy for this problem. Figure 3-6 shows the partial results.

	SolarR	SolarRA	SolarRIA	SolarM	SolarIM
1	NULL	185	185	205	205
2	NULL	185	185	205	205
3	NULL	185	185	205	205
4	NULL	185	185	205	205
5	NULL	185	185	205	205
6	NULL	185	185	205	205
7	NULL	185	185	205	205
8	7	185	7	205	7
9	8	185	8	205	8
10	13	185	13	205	13
11	14	185	14	205	14

Figure 3-6. *Imputing mean and median in the place of NULLs*

T-SQL has enough functions to deal with NULLs. In addition, there are also powerful functions that help you manage strings.

String Operations

There are many string functions in the T-SQL language. This is not a comprehensive overview of them. Please refer to the official SQL Server documentation for the full list. This chapter shows only a few of the useful ones for typical data preparation tasks.

Scalar String Functions

A very frequent data preparation task is string concatenation. In T-SQL, you can use the + string concatenation operator for this task. However, if any of the strings involved in the concatenation is missing or is NULL, the result is NULL as well. You can solve this issue with the ISNULL() function.

However, the CONCAT() function treats NULLs like empty strings, providing a more elegant solution. Also, the CONCAT_WS() function can add a separator between the concatenated strings.

Listing 3-7 shows all of these possibilities on the data from the dbo.vTargetMail view from the AdventureWorksDW2017 demo database.

Listing 3-7. Concatenating Strings

```
SELECT FirstName, MiddleName, LastName,
 FirstName + MiddleName + LastName AS fn1,
 CONCAT(FirstName, MiddleName, LastName) AS fn2,
 CONCAT_WS(', ', FirstName, MiddleName, LastName) AS fn3
FROM dbo.vTargetMail;
```

Figure 3-7 shows the result of the concatenation with the original values.

	FirstName	MiddleName	LastName	fn1	fn2	fn3
1	Jon	V	Yang	JonVYang	JonVYang	Jon, V, Yang
2	Eugene	L	Huang	EugeneLHuang	EugeneLHuang	Eugene, L, Huang
3	Ruben	NULL	Torres	NULL	RubenTorres	Ruben, Torres
4	Christy	NULL	Zhu	NULL	ChristyZhu	Christy, Zhu
5	Elizabeth	NULL	Johnson	NULL	ElizabethJohnson	Elizabeth, Johnson
6	Julio	NULL	Ruiz	NULL	JulioRuiz	Julio, Ruiz
7	Janet	G	Alvarez	JanetGAlvarez	JanetGAlvarez	Janet, G, Alvarez

Figure 3-7. *Concatenated strings*

You might need to check the number of occurrences of a specific character in a string; for example, the number of dots in an e-mail address. You can use the REPLACE() function to replace the searched character with an empty string and the LEN() function to calculate the length of the string before and after the replacement.

Listing 3-8 shows how to count the number of dots in a string.

Listing 3-8. Counting the Number of Dots

```
SELECT
 CONCAT_WS('.', FirstName, MiddleName, AddressLine1) AS s1,
 REPLACE(
  CONCAT_WS('.', FirstName, MiddleName, AddressLine1),
  '.', '') AS s2,
 LEN(CONCAT_WS('.', FirstName, MiddleName, AddressLine1))
  -
 LEN(
  REPLACE(
   CONCAT_WS('.', FirstName, MiddleName, AddressLine1),
   '.', '')
 ) AS nOfDots
FROM dbo.vTargetMail;
```

Figure 3-8 shows the results.

	s1	s2	nOfDots
1	Jon.V.3761 N. 14th St	JonV3761 N 14th St	3
2	Eugene.L.2243 W St.	EugeneL2243 W St	3
3	Ruben.5844 Linden Land	Ruben5844 Linden Land	1
4	Christy.1825 Village Pl.	Christy1825 Village Pl	2
5	Elizabeth.7553 Harness Circle	Elizabeth7553 Harness Circle	1
6	Julio.7305 Humphrey Drive	Julio7305 Humphrey Drive	1
7	Janet.G.2612 Berry Dr	JanetG2612 Berry Dr	2

Figure 3-8. *The number of dots in strings*

The TRANSLATE() function is even more powerful than the REPLACE() function. This function's name is misleading, however; it does not translate a string from one language to another. Instead, it replaces multiple characters at the same time—one set of characters with another. Both sets of characters must have an equal number of elements (or characters). The following code shows how to use the TRANSLATE() function.

```
SELECT CONCAT_WS('. ', EmailAddress, AddressLine1) AS s1,
  TRANSLATE(
   CONCAT_WS('. ', EmailAddress, AddressLine1),
   '@.-', '?/*') AS s2
FROM dbo.vTargetMail;
```

In the code, the @ character is replaced with a ? character, the . is replaced with a /, and the – is replaced with a *.

Aggregating and Splitting Strings

For many years, *aggregating* and *splitting* strings was a complex task in SQL Server. You needed to write complex T-SQL or CLR user-defined functions. The latest versions offer out-of-the-box support for these operations. Listing 3-9 shows the usage of the STRING_AGG() function to get a single string of marital statuses in the education groups from the dbo.vTargetMail view.

Listing 3-9. Aggregating Strings

```
SELECT EnglishEducation,
 STRING_AGG(CAST(MaritalStatus AS NVARCHAR(MAX)), ';')
  AS MSA
FROM dbo.vTargetMail
GROUP BY EnglishEducation;
```

Figure 3-9 shows the aggregated strings.

	EnglishEducation	MSA
1	Bachelors	M;S;M;S;S;S;S;M;S;S;S;M;M;M;S;M;M;S;M;M;M;S;S;S;...
2	Graduate Degree	S;M;S;S;M;M;M;S;S;M;M;S;M;S;M;M;M;S;M;M;S;S;S;S;M;...
3	High School	M;S;S;M;M;M;M;M;S;M;S;S;M;S;S;S;S;S;S;S;S;S;M;M;...
4	Partial College	S;M;M;S;S;S;M;S;M;M;S;S;S;S;S;S;S;M;S;S;S;S;S;M;...
5	Partial High School	M;S;S;M;M;S;M;M;S;S;S;M;M;S;M;S;S;S;S;S;M;S;S;S;S...

Figure 3-9. *Aggregated strings*

The STRING_AGG() function can also sort strings in an aggregated string. Listing 3-10 shows the string aggregation where the source strings are in descending order. Note that the CustomerKey column, which is aggregated, is an integer, and the function performs an implicit conversion.

Listing 3-10. String Aggregation with Order

```
SELECT EnglishEducation,
 STRING_AGG(CustomerKey, ';')
  WITHIN GROUP (ORDER BY CustomerKey DESC) AS CKA
FROM dbo.vTargetMail
WHERE CustomerKey < 11020
GROUP BY EnglishEducation;
```

The result is shown in Figure 3-10.

	EnglishEducation	CKA
1	Bachelors	11014;11013;11012;11011;11010;11009;11008;11007;...
2	High School	11019;11017
3	Partial College	11018;11016;11015

Figure 3-10. *Aggregated sorted strings*

The opposite process of string aggregation is string splitting. This can be done easily with the STRING_SPLIT() function. Listing 3-11 aggregates the strings in a common table expression. This is the query from Listing 3-10. Listing 3-11 uses the STRING_SPLIT() tabular function in the FROM part of the query with the APPLY operator.

Listing 3-11. Splitting Strings

```
WITH CustCTE AS
(
SELECT EnglishEducation,
 STRING_AGG(CustomerKey, ';')
  WITHIN GROUP (ORDER BY CustomerKey DESC) AS CKA
FROM dbo.vTargetMail
WHERE CustomerKey < 11020
GROUP BY EnglishEducation
)
SELECT EnglishEducation, value AS CustomerKey
FROM CustCTE
 CROSS APPLY STRING_SPLIT(CKA, ';')
ORDER BY EnglishEducation DESC;
GO
```

Figure 3-11 shows the result of the string splitting operation.

	EnglishEducation	CustomerKey
1	Partial College	11018
2	Partial College	11016
3	Partial College	11015
4	High School	11019
5	High School	11017
6	Bachelors	11014
7	Bachelors	11013

Figure 3-11. *Strings are split again*

Combining and splitting strings can also be part of adding appropriate derived variables to a dataset. I formally introduce derived variables in the next section, where I also show how you can efficiently group data in T-SQL.

Derived Variables and Grouping Sets

You already learned about the importance of derived variables. For this chapter, I created variables for value imputing NULLs, performing string operations on original variables, and storing the result in a new column. These operations are not overly complex from a T-SQL perspective. The real issue is to find the appropriate derived variables. This is a task where you must know the area you are researching or have a subject matter expert that can help you.

In SQL Server terminology, derived variables are called *computed columns* or *expressions*. The computations can be *row-oriented*, meaning that you add a computed column based on other columns in the same row, or *set-oriented*, when you compute aggregations over groups of rows of the original dataset. Besides simple calculations, I show you how to calculate efficient aggregations in multiple groups in T-SQL.

Adding Computed Columns

The real issue is to find the appropriate derived variables. The following are a few examples.

- Height^2/weight (obesity index)

- Passengers * miles

- Population/area (population density)

- Activation date/ application date (time needed for activation)

- Credit limit/balance

Sometimes you have only a few input variables, and it looks as if there is not much you can do. You should not give up too quickly, however. You can often *extract features* from what initially looks like useless data. The following are a few examples.

- Universal Product Codes (UPCs) store the manufacturer number.

- ZIP codes are in many countries and are mapped geographically.

- You can do many things with dates, such as extracting hierarchies like the year, quarter, or month, finding workdays and weekends, and more.

- Even simple sequential numbers can tell you something; typically, you can conclude (although not with 100% accuracy) that an entity with a higher number was inserted into a database later than an entity with a lower number.

Data normalization and recoding variables from strings to numerics and vice versa can also be treated as creating derived variables. Since these topics are more complex, I devote a section of this chapter to each of them.

For now, let's add a few derived variables to the air quality dataset, as shown in Listing 3-12.

Listing 3-12. Adding Derived Variables

```
SELECT
 ISNULL(SolarR, AVG(SolarR) OVER()) AS SolarRIA,
 IIF(IIF(Ozone IS NULL, 1, 0) = 1 OR
    IIF(SolarR IS NULL, 1, 0) = 1, 1, 0)
```

```
AS NullInRow,
FORMAT(MonthInt, '00') AS MonthChr,
FORMAT(DayInt, '00') AS DayChr,
DATENAME(weekday, '1973' +
 FORMAT(MonthInt, '00') + FORMAT(DayInt, '00'))
AS NameDay,
DATEPART(weekday, '1973' +
 FORMAT(MonthInt, '00') + FORMAT(DayInt, '00'))
AS WeekDay,
CASE
 WHEN DATENAME(weekday, '1973' +
   FORMAT(MonthInt, '00') + FORMAT(DayInt, '00'))
  IN ('Saturday', 'Sunday') THEN 'Weekend'
 ELSE 'Workday'
END AS TypeDay
FROM dbo.airquality;
```

In the code, I input the mean value in the cells with NULLs for the SolarR variable. Next, I added an indicator of whether there is a NULL in a row. Both calculations were introduced earlier in the chapter. Then, I formatted the day and month integers into strings. Finally, I added the name of each day, a number for each day of the week, and the type of day—workday or weekend. Figure 3-12 displays the results.

	SolarRIA	NullInRow	MonthChr	DayChr	NameDay	WeekDay	TypeDay
1	190	0	05	01	Tuesday	3	Workday
2	118	0	05	02	Wednesday	4	Workday
3	149	0	05	03	Thursday	5	Workday
4	313	0	05	04	Friday	6	Workday
5	185	1	05	05	Saturday	7	Weekend
6	185	1	05	06	Sunday	1	Weekend
7	299	0	05	07	Monday	2	Workday

Figure 3-12. *Derived variables*

Data preparation sometimes involves aggregations over groups. Fortunately, SQL Server is strong in this area.

Efficient Grouping

You are likely familiar with the basic GROUP BY clause. Listing 3-13 shows an example.

Listing 3-13. Basic Grouping

```
WITH grCTE AS
(
SELECT Ozone,
 IIF(IIF(Ozone IS NULL, 1, 0) = 1 OR
     IIF(SolarR IS NULL, 1, 0) = 1, 1, 0)
 AS NullInRow,
 CAST( DATEPART(weekday, '1973' +
  FORMAT(MonthInt, '00') + FORMAT(DayInt, '00'))
  AS CHAR(1)) + ' ' +
 DATENAME(weekday, '1973' +
  FORMAT(MonthInt, '00') + FORMAT(DayInt, '00'))
 AS NameDay
FROM dbo.airquality
)
SELECT NameDay, NullInRow,
 SUM(Ozone) AS OzoneTot
FROM grCTE
GROUP BY NameDay, NullInRow
ORDER BY NameDay, NullInRow;
```

In the listing, you see that I first calculated the day of the week and the NULL indicator, and then I grouped over the two variables. There are 14 groups for 14 possible combinations of the seven days and two NULL indicator values. A single aggregation over a combination of both values was done. Figure 3-13 shows the partial results.

NameDay	NullInRow	OzoneTot
1 Sunday	0	736
1 Sunday	1	63
2 Monday	0	533
2 Monday	1	66
3 Tuesday	0	883
3 Tuesday	1	NULL

Figure 3-13. *Basic GROUP BY result*

Imagine that you must do the aggregations over each of the two grouping variables separately in the previous query. With the basic GROUP BY clause, you would need to write two SELECT statements. This is where the GROUPING SETS clause becomes handy. You can specify multiple groupings in a single statement. This way, SQL Server also has more options for optimization than with two separate statements; therefore, a single query could execute faster than two separate ones. Listing 3-14 shows an example of a single statement with multiple groupings.

Listing 3-14. Using GROUPING SETS

```
WITH grCTE AS
(
SELECT Ozone,
 IIF(IIF(Ozone IS NULL, 1, 0) = 1 OR
     IIF(SolarR IS NULL, 1, 0) = 1, 1, 0)
 AS NullInRow,
 CAST( DATEPART(weekday, '1973' +
  FORMAT(MonthInt, '00') + FORMAT(DayInt, '00'))
  AS CHAR(1)) + ' ' +
 DATENAME(weekday, '1973' +
  FORMAT(MonthInt, '00') + FORMAT(DayInt, '00'))
 AS NameDay
FROM dbo.airquality
)
SELECT NameDay, NullInRow,
 SUM(Ozone) AS OzoneTot
FROM grCTE
```

```
GROUP BY GROUPING SETS
 (NameDay, NullInRow)
ORDER BY NameDay, NullInRow;
```

The result consists of nine rows: seven for the names of the days and two for the values of the NULL indicator, as seen in Figure 3-14.

	NameDay	NullInRow	OzoneTot
1	NULL	0	4673
2	NULL	1	214
3	1 Sunday	NULL	799
4	2 Monday	NULL	599
5	3 Tuesday	NULL	883
6	4 Wednesday	NULL	759
7	5 Thursday	NULL	618
8	6 Friday	NULL	474
9	7 Saturday	NULL	755

Figure 3-14. *Two groupings in a single result set*

But what if you needed aggregations independently over each grouping variable, over both combined, and aggregation over the whole dataset, without grouping? You can use the CUBE clause for this task, as seen in Listing 3-15.

Listing 3-15. Using the CUBE Clause

```
WITH grCTE AS
(
SELECT Ozone,
 IIF(IIF(Ozone IS NULL, 1, 0) = 1 OR
     IIF(SolarR IS NULL, 1, 0) = 1, 1, 0)
 AS NullInRow,
 CAST( DATEPART(weekday, '1973' +
  FORMAT(MonthInt, '00') + FORMAT(DayInt, '00'))
  AS CHAR(1)) + ' ' +
 DATENAME(weekday, '1973' +
  FORMAT(MonthInt, '00') + FORMAT(DayInt, '00'))
 AS NameDay
```

```
FROM dbo.airquality
)
SELECT NameDay, NullInRow,
 SUM(Ozone) AS OzoneTot
FROM grCTE
GROUP BY CUBE
 (NameDay, NullInRow)
ORDER BY NameDay, NullInRow;
```

Figure 3-15 shows the partial results.

NameDay	NullInRow	OzoneTot
NULL	NULL	4887
NULL	0	4673
NULL	1	214
1 Sunday	NULL	799
1 Sunday	0	736
1 Sunday	1	63
2 Monday	NULL	599
2 Monday	0	533
2 Monday	1	66

Figure 3-15. *All possible groupings over two variables*

Figure 3-15 shows that NULLs are used when there are *hyper-aggregates* in the output. Basic grouping is over the two input variables combined, so both are known; for example, there is a grouping for NameDay equal to Sunday and NullInRow equal to 1. Then there are hyper-aggregates over the NameDay only, where NullInRow makes no sense; for example, aggregate Monday, NULL. Then there are hyper-aggregates over NullInRow variable only, where NameDay makes no sense; for example, aggregate NULL, 1. And finally, there is a hyper-aggregate over the whole dataset; this is the row where both grouping variables are NULL in the result. The result set has 24 rows altogether: 14 for aggregations over combinations of both input variables, 7 for aggregations over days, 2 for aggregations over the NULL indicator, and 1 for the aggregation over the whole dataset.

There are more options with the GROUPING SETS clause. To fully master it, please refer to the SQL Server documentation and check the ROLLUP clause and the GROUPING() and GROUPING_ID() functions.

Data Normalization

Data normalization brings variables to the same scale, so you have comparable values. In addition, you can control possible *outliers*. Outliers are rare out-of-bound values that can greatly impact the statistical calculations you are performing. For example, a single extremely large value can substantially change a variable's mean.

Range and Z-score Normalization

With *range normalization*, you get all values in the range between 0 and 1, with the distribution that follows the distribution of the original variables. Range normalization is a *linear* transformation. The following is the formula for the range normalization.

$$V = \frac{x - x_{min}}{x_{max} - x_{min}}$$

The range normalization is very simple and intuitive. However, with the range normalization, you are not limiting the influence of the outliers. There might also be a problem with new data. If you get a new maximal or a new minimal value in the data, you must recalculate all previous range normalization values. This problem is solved with the *Z-score normalization*. With this normalization, you calculate the Z-score for every value of the input variable. You subtract the mean values from every single value and then divide the result with the standard deviation for the variable, as the following formula shows.

$$V = \frac{x - \mu}{\sigma}$$

After the Z-score normalization, the mean of the new variable is zero, and the standard deviation is one. Listing 3-16 shows the range and the Z-score normalizations.

Listing 3-16. Range and Z-score Normalizations

```
WITH castCTE AS
(
SELECT MonthInt, DayInt,
 CAST(Wind AS NUMERIC(8,2)) AS Wind,
 CAST(SolarR AS NUMERIC(8,2)) AS SolarR
```

```
FROM dbo.airquality
)
SELECT MonthInt, DayInt,
 CAST(ROUND(
 (Wind - MIN(Wind) OVER()) /
  (MAX(Wind) OVER() - MIN(Wind) OVER())
 , 2) AS NUMERIC(8,2)) AS WindR,
 CAST(ROUND(
 (SolarR - MIN(SolarR) OVER()) /
  (MAX(SolarR) OVER() - MIN(SolarR) OVER())
 , 2) AS NUMERIC(8,2)) AS SolarRR,
 ROUND(
 (Wind - AVG(Wind) OVER()) /
  STDEV(Wind) OVER()
 , 2) AS WindS,
 ROUND(
 (SolarR - AVG(SolarR) OVER()) /
  STDEV(SolarR) OVER()
 , 2) AS SolarRS
FROM castCTE
ORDER BY MonthInt, DayInt;
```

In Listing 3-16, in the top common table expression, I first cast the two input variables to numeric data type to avoid integer calculations later in the query. The result is shown in Figure 3-16.

	MonthInt	DayInt	WindR	SolarRR	WindS	SolarRS
1	5	1	0.32	0.56	-0.71	0.05
2	5	2	0.37	0.34	-0.42	-0.75
3	5	3	0.58	0.43	0.72	-0.41
4	5	4	0.53	0.94	0.44	1.41
5	5	5	0.68	NULL	1.3	NULL
6	5	6	0.68	NULL	1.3	NULL
7	5	7	0.37	0.89	-0.42	1.26

Figure 3-16. *The result of the range and Z-score normalizations*

Let's also check the mean and the standard deviation for the normalized variables in Listing 3-17.

Listing 3-17. Calculating the First Two Moments After the Normalization

```
WITH castCTE AS
(
SELECT MonthInt, DayInt,
 CAST(Wind AS NUMERIC(8,2)) AS Wind,
 CAST(SolarR AS NUMERIC(8,2)) AS SolarR
FROM dbo.airquality
),
normCTE AS (
SELECT MonthInt, DayInt,
 CAST(ROUND(
 (Wind - MIN(Wind) OVER()) /
  (MAX(Wind) OVER() - MIN(Wind) OVER())
 , 2) AS NUMERIC(8,2)) AS WindR,
 CAST(ROUND(
 (SolarR - MIN(SolarR) OVER()) /
  (MAX(SolarR) OVER() - MIN(SolarR) OVER())
 , 2) AS NUMERIC(8,2)) AS SolarRR,
 ROUND(
 (Wind - AVG(Wind) OVER()) /
  STDEV(Wind) OVER()
 , 2) AS WindS,
 ROUND(
 (SolarR - AVG(SolarR) OVER()) /
  STDEV(SolarR) OVER()
 , 2) AS SolarRS
FROM castCTE
)
SELECT 'Range' AS normType,
 ROUND(AVG(WindR), 4) AS avgWR,
 ROUND(STDEV(WindR), 4) AS stdevWR,
 ROUND(AVG(SolarRR), 4) AS avgSR,
```

```
 ROUND(STDEV(SolarRR), 4) AS stdevSR
FROM normCTE
UNION ALL
SELECT 'Z-score' AS normType,
 ROUND(AVG(WindS), 4) AS avgWS,
 ROUND(STDEV(WindS), 4) AS stdevWS,
 ROUND(AVG(SolarRS), 4) AS avgSS,
 ROUND(STDEV(SolarRS), 4) AS stdevSS
FROM normCTE
ORDER BY normType;
```

Figure 3-17 shows the results.

	normType	avgWR	stdevWR	avgSR	stdevSR
1	Range	0.4458	0.1833	0.5473	0.2755
2	Z-score	0.0002	1.0007	0.0004	0.9998

Figure 3-17. *The first two moments after the normalization*

After the Z-score normalization, the mean is (up to three decimal numbers) at zero, and the standard deviation is at one.

Logistic and Hyperbolic Tangent Normalization

Logistic and *hyperbolic tangent normalizations* are both non-linear transformations. You transform the values of the original variable into new values with a non-linear function. Let's go straight to the formula for logistic normalization.

$$V = \frac{1}{1 + e^{-x}}$$

From the formula, you can realize that the e^{-x} is the non-linear part. The logistic function transforms data into the *interval between 0 and +1*. The function has the typical S-shape. It is nearly linear when the source variable is around zero and asymptotically approaches the value 0 on the left side and the value 1 on the right side, where the values of the original variable are very small or very big. This way, the impact of the extreme absolute values is diminished, yet you can still operate with all the values. You can see the logistic function compared to the linear function in Figure 3-18. Please note that there are two Y axes. The one on the left is for the linear function and is between –6 and +6. The one on the right is for the logistic function and is between 0 and +1. If I used the same scale for both functions, the logistic function would be flattened, and it would not be that easy to see the S-shape.

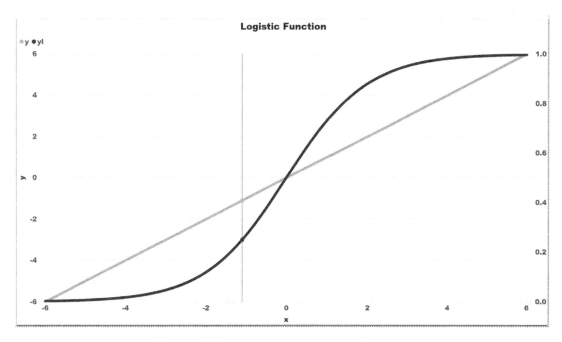

Figure 3-18. *The logistic function*

The hyperbolic tangent function has a similar S-shape as the logistic function. However, it transforms the data into an *interval between –1 and +1*. The following is the formula.

$$V = \frac{e^x - e^{-x}}{e^x + e^{-x}}$$

You can see the hyperbolic tangent function in Figure 3-19. Please note that the values in the right Y axis go from –1 to +1. I used two Y axes for the same reason as with the logistic function.

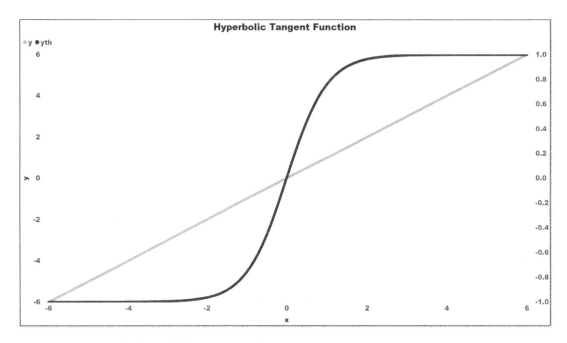

Figure 3-19. *The hyperbolic tangent function*

You should feed both logistic and hyperbolic tangent functions with numeric data that is centered around zero. If you have only positive values, most of them greater than 2, the majority of the values after both transformations are very close to +1. Listing 3-18 shows centering the Wind variable around 0 by subtracting the mean from each value and then transforming this centered value with the logistic and hyperbolic tangent normalizations.

Listing 3-18. Logistic and Hyperbolic Tangent Normalizations

```
SELECT Wind,
 Wind - AVG(Wind) OVER() AS WindC,
 1.0 / (1.0 + EXP(-(Wind - AVG(Wind) OVER())))
  AS WindNLogistic,
 (EXP(Wind - AVG(Wind) OVER()) - EXP(-(Wind - AVG(Wind) OVER())))
 /
 (EXP(Wind - AVG(Wind) OVER()) + EXP(-(Wind - AVG(Wind) OVER())))
  AS WindNTanH
FROM dbo.airquality;
```

Figure 3-20 shows the partial result of Listing 3-18.

	Wind	WindC	WindNLogistic	WindNTanH
1	5	-4	0.0179862099620916	-0.999329299739067
2	9	0	0.5	0
3	7	-2	0.119202922022118	-0.964027580075817
4	11	2	0.880797077977882	0.964027580075817
5	9	0	0.5	0
6	10	1	0.731058578630005	0.761594155955765
7	8	-1	0.268941421369995	-0.761594155955765

Figure 3-20. *The result of the logistic and hyperbolic tangent normalizations*

Note   The query in Listing 3-18 does not include the ORDER BY clause. Therefore, you can get a different order of rows when you execute the query. I intentionally omitted the ORDER BY clause to show some of the smaller and the bigger variables' values in the shortened output.

If you want to operate with positive values only, you should use the logistic normalization; if you want to have negative values after the normalization, you should use the hyperbolic tangent transformation.

Recoding Variables

A very frequent data preparation task is *variable recoding*. Sometimes you want to use string variables in a regression algorithm that expects numbers only. The opposite can happen as well—maybe you want to use the Bayesian inference algorithm for predicting the states of a discrete variable, and you have a continuous variable that you would like to use for the input. You need to discretize or bin the numeric variable into a few predefined classes or bins in such a case.

Converting Strings to Numerics

When converting string variables to numerics, you have two possibilities: the character variables can be either ordinal or nominal. Ordinal variables have intrinsic order. You use this order to generate appropriate numerical values. Listing 3-19 shows how you can do this with the CASE expression.

This example returns to the mtcars dataset and recodes the hpdescription variable.

Listing 3-19. Recoding an Ordinal Variable

```
SELECT carbrand,
 hp, hpdescription,
 CASE hpdescription
  WHEN N'Weak' THEN 1
  WHEN N'Medium' THEN 2
  WHEN N'Strong' THEN 3
 END AS hpdescriptionint
FROM dbo.mtcars;
```

The result of the recoding process is shown in Figure 3-21.

	carbrand	hp	hpdescription	hpdescriptionint
1	AMC Javelin	150	Medium	2
2	Cadillac Fleetwood	205	Strong	3
3	Camaro Z28	245	Strong	3
4	Chrysler Imperial	230	Strong	3
5	Datsun 710	93	Weak	1
6	Dodge Challenger	150	Medium	2
7	Duster 360	245	Strong	3

Figure 3-21. *Recoded variable*

Let's check the distribution of the values after the recoding. Listing 3-20 shows the query that calculates the frequencies of the recoded variable.

Listing 3-20. Calculating Frequencies

```
-- Checking the distribution
WITH frequency AS
(
SELECT
 CASE hpdescription
```

```
        WHEN N'Weak' THEN 1
        WHEN N'Medium' THEN 2
        WHEN N'Strong' THEN 3
     END AS hpdescriptionint,
 COUNT(hpdescription) AS AbsFreq,
 CAST(ROUND(100. * (COUNT(hpdescription)) /
        (SELECT COUNT(*) FROM dbo.mtcars), 0) AS int) AS AbsPerc
FROM dbo.mtcars AS v
GROUP BY v.hpdescription
)
SELECT hpdescriptionint,
 AbsFreq,
 SUM(AbsFreq)
  OVER(ORDER BY hpdescriptionint
        ROWS BETWEEN UNBOUNDED PRECEDING
        AND CURRENT ROW) AS CumFreq,
 AbsPerc,
 SUM(AbsPerc)
  OVER(ORDER BY hpdescriptionint
        ROWS BETWEEN UNBOUNDED PRECEDING
        AND CURRENT ROW) AS CumPerc,
 CAST(REPLICATE('*', AbsPerc) AS varchar(50)) AS Histogram
FROM frequency
ORDER BY hpdescriptionint;
```

Figure 3-22 shows the results.

	hpdescriptionint	AbsFreq	CumFreq	AbsPerc	CumPerc	Histogram
1	1	9	9	28	28	****************************
2	2	13	22	41	69	***
3	3	10	32	31	100	*******************************

Figure 3-22. *Frequencies of the recoded variable*

With nominal variables, you cannot use the order because there is no order in the values of the variable. What you can do is create indicators, called dummy variables or *dummies*. You create one new variable for each possible state of the original nominal variable.

When a specific state is taken, the appropriate dummy takes value 1; otherwise, it takes value 0. In every single row, only one dummy variable can take the value 1. Listing 3-21 shows how you can create dummies on the *engine* variable.

Listing 3-21. Creating Dummies

```
SELECT carbrand, engine,
 IIF(engine = 'V-shape', 1, 0)
  AS engineV,
 IIF(engine = 'Straight', 1, 0)
  AS engineS
FROM dbo.mtcars;
```

Figure 3-23 shows the result.

	carbrand	engine	engineV	engineS
1	AMC Javelin	V-shape	1	0
2	Cadillac Fleetwood	V-shape	1	0
3	Camaro Z28	V-shape	1	0
4	Chrysler Imperial	V-shape	1	0
5	Datsun 710	Straight	0	1
6	Dodge Challenger	V-shape	1	0
7	Duster 360	V-shape	1	0

Figure 3-23. *Dummies from the engine variable*

When you use dummies in further analysis, you should omit at least one of them. This is connected with the number of degrees of freedom in the original variable, which is the number of distinct states minus one. There is a perfect association between all dummies. For example, if the original variable has four states, you need only three dummies to perfectly predict or determine the fourth one.

Discretizing Numerical Variables

The last task in this chapter is the *discretization of numeric variables*. Discretization itself is an important area in statistics and data science. The following are the three simplest and most popular methods, which cover the majority of needs.

- **Equal width discretization.** The new classes of the discrete variable have equal width but a different number of cases in each class.

- **Equal height discretization.** The new classes of the discrete variable have an equal height of frequency bars, meaning (approximately) an equal number of cases in each class. However, the width of the classes varies.

- **Custom discretization.** You define the new classes based on the understanding of the data and the business problem. Classes have neither equal width nor equal height.

For equal width binning, you need to determine the width of the bins. For this, you need to find the minimal and the maximal value of the numeric variable, calculate the range, and divide the range with the number of bins you want to have after the discretization. The following query checks this info for the hp variable of the dbo.mtcars table.

```
SELECT MIN(hp) AS minA,
 MAX(hp) AS maxA,
 MAX(hp) - MIN(hp) AS rngA,
 AVG(hp) AS avgA,
 1.0 * (MAX(hp) - MIN(hp)) / 3 AS binwidth
FROM dbo.mtcars;
```

The results of this query are not shown. I only show the query to help you better understand the code that does the equal width binning, as seen in Listing 3-22.

Listing 3-22. Equal Width Binning

```
DECLARE @binwidth AS NUMERIC(5,2),
 @minA AS INT, @maxA AS INT;
SELECT @minA = MIN(hp),
 @maxa = MAX(hp),
 @binwidth = 1.0 * (MAX(hp) - MIN(hp)) / 3
FROM dbo.mtcars;
SELECT carbrand, hp,
```

```
CASE
  WHEN hp >= @minA + 0 * @binwidth AND hp < @minA + 1 * @binwidth
    THEN CAST((@minA + 0 * @binwidth) AS VARCHAR(10)) + ' - ' +
         CAST((@minA + 1 * @binwidth - 1) AS VARCHAR(10))
  WHEN hp >= @minA + 1 * @binwidth AND hp < @minA + 2 * @binwidth
    THEN CAST((@minA + 1 * @binwidth) AS VARCHAR(10)) + ' - ' +
         CAST((@minA + 2 * @binwidth - 1) AS VARCHAR(10))
  ELSE CAST((@minA + 2 * @binwidth) AS VARCHAR(10)) + ' + '
 END AS hpEWB
FROM dbo.mtcars
ORDER BY carbrand;
```

I used the bin width stored in the variable @binwidth in the CASE expression to calculate the limits of the intervals for the three bins.

Listing 3-23 shows the equal height binning, which is simple in T-SQL with the NTILE() window ranking function.

Listing 3-23. Equal Height Binning

```
SELECT carbrand, hp,
 CAST(NTILE(3) OVER(ORDER BY hp)
  AS CHAR(1)) AS hpEHB
FROM dbo.mtcars
ORDER BY carbrand;
```

Finally, Listing 3-24 shows the custom binning. This time I use nested IIF() expressions, because I need only three bins.

Listing 3-24. Custom Binning

```
SELECT carbrand, hp,
 IIF(hp > 175, '3 - Strong',
     IIF (hp < 100, '1 - Weak', '2 - Medium'))
  AS hpCUB
FROM dbo.mtcars
ORDER BY carbrand;
```

Why didn't I show the results of the three binning processes yet? I used the previous three queries only to show each binning separately for better understanding. Listing 3-25 shows all three binning's together, in a single piece of code.

Listing 3-25. The Three Different Binning Options

```
DECLARE @binwidth AS NUMERIC(5,2),
 @minA AS INT, @maxA AS INT;
SELECT @minA = MIN(hp),
 @maxa = MAX(hp),
 @binwidth = 1.0 * (MAX(hp) - MIN(hp)) / 3
FROM dbo.mtcars;
SELECT carbrand, hp,
 CASE
  WHEN hp >= @minA + 0 * @binwidth AND hp < @minA + 1 * @binwidth
   THEN CAST((@minA + 0 * @binwidth) AS VARCHAR(10)) + ' - ' +
       CAST((@minA + 1 * @binwidth - 1) AS VARCHAR(10))
  WHEN hp >= @minA + 1 * @binwidth AND hp < @minA + 2 * @binwidth
   THEN CAST((@minA + 1 * @binwidth) AS VARCHAR(10)) + ' - ' +
       CAST((@minA + 2 * @binwidth - 1) AS VARCHAR(10))
  ELSE CAST((@minA + 2 * @binwidth) AS VARCHAR(10)) + ' + '
 END AS hpEWB,
 CHAR(64 + (NTILE(3) OVER(ORDER BY hp)))
  AS hpEHB,
 IIF(hp > 175, '3 - Strong',
     IIF (hp < 100, '1 - Weak', '2 - Medium'))
  AS hpCUB
FROM dbo.mtcars
ORDER BY carbrand;
```

You can see the result in Figure 3-24.

	carbrand	hp	hpEWB	hpEHB	hpCUB
1	AMC Javelin	150	146.33 - 239.66	B	2 - Medium
2	Cadillac Fleetwood	205	146.33 - 239.66	C	3 - Strong
3	Camaro Z28	245	240.66 +	C	3 - Strong
4	Chrysler Imperial	230	146.33 - 239.66	C	3 - Strong
5	Datsun 710	93	52.00 - 145.33	A	1 - Weak
6	Dodge Challenger	150	146.33 - 239.66	B	2 - Medium
7	Duster 360	245	240.66 +	C	3 - Strong
8	Ferrari Dino	175	146.33 - 239.66	B	2 - Medium
9	Fiat 128	66	52.00 - 145.33	A	1 - Weak

Figure 3-24. *The hp variable binned in three different ways*

Figure 3-25 is a graphical representation of the three new binned variables.

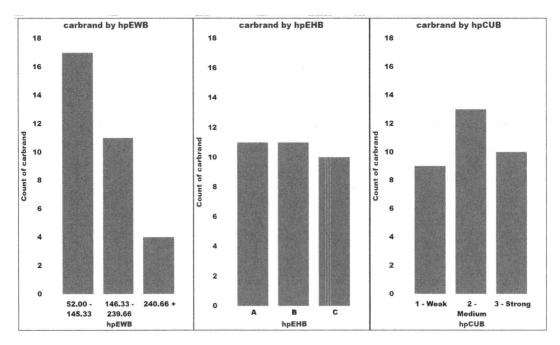

Figure 3-25. *The distribution of the three binned variables*

When do you use which discretization method? With equal width binning, you are preserving the shape of the distribution of the original variable, especially if you use enough bins. This is very useful for data overview when you want to see the distribution of a continuous variable with the same graph type as the distribution of discrete variables.

With equal height binning, you are maximally preserving the variability or the *information* that the variable holds. When you lower the number of possible states, you are losing information; with equal height binning, the loss is the minimal possible. I return to this topic in Chapter 4, which deals with data quality. There, I introduce *entropy*, which measures the amount of information in a variable. You can use equal height binning when you want to use the variable further in analysis, and you do not see a better option with custom binning.

Very often, custom binning, binning from the business perspective, is the best option. The hp variable from the mtcars dataset is a good example. The variable is measuring the power of an engine. When the power is small, a difference of 5 hp; for example, the difference between 50 hp and 55 hp is important. However, the difference between 300 hp and 305 hp is negligible. Therefore, it makes sense to make narrower classes for small values and wider classes for big values.

Conclusion

Data preparation is extremely important for a successful analysis. It is a never-ending story. You can do the preparation and perform the analysis, but you are not satisfied with the results. You can go back to the data preparation stage, find a few more derived variables, discretize few variables differently, and do the analysis again.

However, even the best data preparation can't help you if the quality of the data is poor. Unfortunately, this is often a serious issue with data that you must deal with in real life. I discuss how to find problematic data and measure data quality in the next chapter.

CHAPTER 4

Data Quality and Information

You start a shiny new analytical project. You do the initial overview of the data. And bang! You meet probably the biggest issue in advanced analysis and business intelligence: *data quality*. Garbage in, garbage out is a very old rule. Before doing advanced analyses, it is always recommended that you check the data quality. Measuring improvements in data quality over time can help to understand the factors that influence it.

Measuring data quality is tightly connected with or is nearly a synonym for *data profiling*. Data profiling helps you find areas with low data quality. You measure different *aspects* of data quality. These aspects are called *data quality dimensions*. Please note that data quality dimensions have nothing to do with data warehouse dimensions. Again, data quality dimensions are aspects of data quality. The following are some typical aspects.

- *Completeness*, which is the amount of data that is known and the amount that is unknown

- *Accuracy*, which is how correct the data is

- *Information*, which is whether the data is useful for analyses

The dimensions are also called *hard dimensions*. Hard dimensions can be measured programmatically, such as with T-SQL queries. In the data quality theory, *soft dimensions* are also defined. Soft dimensions are about users' perception of the data, and you can research them by interviewing the users. For example, you can ask users whether they trust the data they get from the database. This book focuses on T-SQL language, not on the data quality; therefore, I will not spend more time on the soft dimensions.

The information data quality dimension is extremely important for analytical projects. Measuring the information involves a bit more complex queries than measuring completeness and accuracy. Because of that, I promoted information in the

© Dejan Sarka 2021
D. Sarka, *Advanced Analytics with Transact-SQL*, https://doi.org/10.1007/978-1-4842-7173-5_4

chapter title, and I spend a large portion of this chapter dealing with measuring the information in the data.

Data Quality

I used the two datasets I imported from R, stored them in the dbo.mtcars and dbo.airquality tables, which are in the AdventureWorksDW2017 demo database. I added a few rows with data of low quality. I created two new views from the base tables, as Listing 4-1 shows.

Listing 4-1. Creating View with Bad Data

```
USE AdventureWorksDW2017;
GO
DROP VIEW IF EXISTS dbo.airqualityLQ;
GO
CREATE OR ALTER VIEW dbo.airqualityLQ
AS
SELECT Ozone, SolarR, Wind, Temp,
  CAST('1973' + FORMAT(MonthInt, '00') +
       FORMAT(DayInt, '00') AS date) AS DateM
FROM dbo.airquality
UNION
SELECT *
FROM (VALUES
 (150, NULL, 55,     -- too strong Wind
  88, '19731001'),  -- too high temperature, date
 (80, 120, 8,
  92, '19731002'),  -- too high temperature, date
 (50, 40, 6,
  56, '19730202')   -- too low date
 ) AS a(Ozone, SolarR, Wind,
  Temp, DateM);
GO
DROP VIEW IF EXISTS dbo.mtcarsLQ;
GO
```

```
CREATE OR ALTER VIEW dbo.mtcarsLQ
AS
SELECT carbrand, engine,
 transmission + '@comp.com' AS transMail
FROM dbo.mtcars
UNION
SELECT *
FROM (VALUES
  (N'1',                              -- Short carbrand
   N'Straight',
   N'Manual#comp.com'),              -- Incorrect mail
  (N'2 Long Carbrand Name' +
   REPLICATE(N'X',100),              -- Long carbrand
   N'V-shape',
   N'Manual@comp.com'),
  (N'3 Normal Carbrand',
   N'Streight',                       -- Spelling error
   N'Manual#comp.com')               -- Incorrect mail
  ) AS a(carbrand, engine, transMail);
GO
```

You can see the errors I made from the code and the comments in Listing 4-1. I added fictitious mail addresses to the data from the dbo.mtcars table. You can check the content of the two new views with the following code.

```
SELECT *
FROM dbo.airqualityLQ
ORDER BY dateM;
SELECT *
FROM dbo.mtcarsLQ
ORDER BY carbrand;
```

Now let's check the completeness and the accuracy of the data.

Measuring Completeness

Measuring completeness involves counting the number of NULLs in a column and the number of rows with NULLs in a table. I introduced all the queries needed in Chapter 3. The same queries are used here, but this time I queried the dbo.airqualityLQ view instead of the base table, as shown in Listing 4-2.

Listing 4-2. Counting NULLs in Columns

```
SELECT IIF(Ozone IS NULL, 1, 0) AS OzoneN,
 COUNT(*) AS cnt
FROM dbo.airquality
GROUP BY IIF(Ozone IS NULL, 1, 0);
SELECT IIF(SolarR IS NULL, 1, 0) AS SolarRN,
 COUNT(*) AS cnt
FROM dbo.airqualityLQ
GROUP BY IIF(SolarR IS NULL, 1, 0);
```

Figure 4-1 shows the results.

	OzoneN	cnt
1	0	116
2	1	37

	SolarRN	cnt
1	0	148
2	1	8

Figure 4-1. *Number of missing values in columns*

There are 37 NULLs in the OzoneN column and 8 in the SolarRN column. With the help of the IIF() function, you can measure the completeness of the table (or the view) and the number of rows without any NULLs in a column, as shown in Listing 4-3.

Listing 4-3. Checking the Completeness of a View

```
SELECT
 IIF(IIF(Ozone IS NULL, 1, 0) = 1 OR
     IIF(SolarR IS NULL, 1, 0) = 1, 1, 0)
 AS NullInRow,
 COUNT(*) AS cnt
FROM dbo.airqualityLQ
GROUP BY
 IIF(IIF(Ozone IS NULL, 1, 0) = 1 OR
     IIF(SolarR IS NULL, 1, 0) = 1, 1, 0);
```

In Figure 4-2, there are 113 complete rows and 43 rows with NULLs.

	NullInRow	cnt
1	0	113
2	1	43

Figure 4-2. *Table (or view) completeness*

Measuring completeness was simple. However, it can become more complex if you need to search for the missing rows. For example, in *time series* data, you have data in different time points, and many times you cannot allow a missing time point. I introduce time series data in Chapter 5. Finding inaccurate data is always more complex than measuring completeness. Data can be inaccurate in many ways.

Finding Inaccurate Data

The problem with the accuracy is that you can find *suspicious* data, which is not always inaccurate. Sometimes things are simple. For example, in your database, you might have the birth date for an existing customer equal to December 12, 1770. This is clearly an error. Nobody can be 250 years old. But what about December 12, 1915? It is possible that a customer is slightly older than 105 years, but an error is also possible. When you find suspicious data, you must check the values with the subject area experts before you define the suspicious value as erroneous.

Data can be inaccurate in many ways. In a relational database, one of the first things you check when you import data from other sources whether the rows are *unique*. It is a good practice that every table in a SQL Server database has a *primary key*. Columns that are candidates for the primary key must be unique and known (not NULL). The query in Listing 4-4 checks for the uniqueness of the carbrand column.

Listing 4-4. Checking for the Uniqueness

```
SELECT carbrand,
 COUNT(*) AS Number
FROM dbo.mtcarsLQ
GROUP BY carbrand
HAVING COUNT(*) > 1
ORDER BY Number DESC;
```

The query did not return any rows; therefore, the values in this column are unique, so the column is a good candidate for the primary key.

You can check the accuracy of the values of a continuous numeric variable by checking the extremes. For example, you can search for values that are more than two standard deviations away from the mean. In a normal distribution, values that are more than two standard deviations away from the mean count for less than 5% of all values. The area under each tail of the distribution function is less than 2.5% of the total area, as you might remember from Figure 2-1 from Chapter 2. Listing 4-5 checks for extremely high values in the Wind variable. Since the variable cannot take a value lower than zero, it makes no sense to check the values under the left tail of the distribution.

Listing 4-5. Checking for the Extreme Numerical Values

```
WITH WindCTE AS
(
SELECT DateM, Wind,
 AVG(Wind) OVER() AS WindAvg,
 STDEV(Wind) OVER() AS WindStDev
FROM dbo.airqualityLQ
)
SELECT DateM, Wind,
 WindAvg, WindStDev
```

```
FROM WindCTE
WHERE Wind >
 WindAvg + 2 * WindStDev
ORDER BY DateM;
```

Figure 4-3 shows the result. The first two rows show a value of 20, which is not that extreme and is probably not an error. However, the 55 value for the Wind variable in the third row is farther away from the mean and potentially erroneous.

	DateM	Wind	WindAvg	WindStDev
1	1973-05-09	20	9	5.03422118714751
2	1973-06-17	20	9	5.03422118714751
3	1973-10-01	55	9	5.03422118714751

Figure 4-3. *Suspicious numeric values*

You can check the dates of the extreme values. I checked the number of rows for each month in the air quality data, which is shown in Listing 4-6.

Listing 4-6. Checking Dates

```
SELECT MONTH(DateM) AS DateMonth,
 COUNT(*) AS DaysInMonth
FROM dbo.airqualityLQ
GROUP BY MONTH(DateM)
ORDER BY DaysInMonth;
```

Figure 4-4 shows the complete result.

	DateMonth	DaysInMonth
1	2	1
2	10	2
3	9	30
4	6	30
5	7	31
6	8	31
7	5	31

Figure 4-4. *Suspicious months*

113

It is easy to see that the dates in February (the month equal to 2) and October (10) are rare. The single row with a date in February is suspicious because the second date is in May, and there are no rows for March or April. The two rows in October might be correct because there are rows for every day in September.

Strings are less constrained than numbers or dates and thus are harder to check for accuracy. If a variable is not discrete and can take a value from an unlimited pool, the situation is worse. Any value could be correct.

You can get clues about suspicious values by checking the distribution of string lengths. Listing 4-7 checks the carbrand variable.

Listing 4-7. Checking the Length of the Strings

```
SELECT LEN(carbrand) AS carbrandLength,
 COUNT(*) AS Number
FROM dbo.mtcarsLQ
GROUP BY LEN(carbrand)
ORDER BY Number, carbrandLength;
```

The results in Figure 4-5 show that there are two suspicious values.

	carbrandLength	Number
1	1	1
2	7	1
3	18	1
4	19	1
5	120	1
6	12	2
7	16	2
8	8	3

Figure 4-5. *Suspicious lengths of the car brand*

Look at the number of rows with the length of the car brand equal to 1 and equal to 120. In both cases, there is a single row for this length. There is also a single row for lengths 7, 18, and 19. However, these lengths look much more plausible that lengths 1 and 120. Listing 4-8 finds those two suspicious values.

Listing 4-8. Finding the Values with Suspicious Length

```
SELECT carbrand
FROM dbo.mtcarsLQ
WHERE LEN(carbrand) < 2
   OR LEN(carbrand) > 20;
```

The two rows are shown in Figure 4-6.

	carbrand
1	1
2	2 Long Carbrand NameXXXXXXXXXXXXXXXXXXXXXXX...

Figure 4-6. *Suspicious car brands*

If a variable is discrete, you know that the pool of the possible values is limited. You can check the distribution of the values, as shown in Listing 4-9.

Listing 4-9. Checking the Distribution of Values

```
SELECT engine,
 COUNT(*) AS Number
FROM dbo.mtcarsLQ
GROUP BY engine
ORDER BY Number;
```

Figure 4-7 shows the results. The "Streight" value has a count of only 1. This is probably a spelling error; you can safely guess that the value should be "Straight".

	engine	Number
1	Streight	1
2	Straight	15
3	V-shape	19

Figure 4-7. *A suspicious value has very low frequency*

Sometimes strings must comply with a pattern. For example, email addresses follow strict rules. You can check whether the strings comply with allowed patterns with *regular expressions*. Unfortunately, you cannot validate a string against a regular expression in pure T-SQL. In T-SQL, you can use the LIKE operator, which allows checking for simple patterns only. Listing 4-10 checks email addresses with the LIKE operator.

Listing 4-10. Checking Patterns

```
SELECT carbrand, transMail
FROM dbo.mtcarsLQ
WHERE transMail NOT LIKE '%@%';
```

The result is shown in Figure 4-8.

	carbrand	transMail
1	1	Manual#comp.com
2	3 Normal Carbrand	Manual#comp.com

Figure 4-8. *Irregular email addresses*

You can bring regular expressions to T-SQL through another language that you can run in the database engine. SQL Server supports CLR integration, allowing you to use Visual C# or Visual Basic and external languages, including R, Python, and Java. So far in this book, I used T-SQL code in R only, so I show an example in R. For this, I needed to download the data.table R package. I ran the following R code in RStudio IDE, the most popular development environment for R, to get the data.table package.

```
download.packages("data.table", destdir="C:\\Apress\\Ch04",
                  type="win.binary")
```

Next, I loaded the package to my SQL Server database. This is possible with the CREATE EXTERNAL LIBRARY T-SQL command, which shows the following code.

```
CREATE EXTERNAL LIBRARY [data.table]
FROM (CONTENT = 'C:\Apress\Ch04\data.table_1.12.0.zip')
WITH (LANGUAGE = 'R');
```

The sys.sp_execute_external_script procedure can be used to find the rows that do not comply with the regular expression for the email addresses, as Listing 4-11 shows.

Listing 4-11. Validating Email Addresses Against a Regular Expression

```
EXECUTE sys.sp_execute_external_script
 @language=N'R',
 @script =
  N'
    library(data.table)
    RTD <- data.table(RT)
    regex <- "^[[:alnum:]._-]+@[[:alnum:].-]+$"
    RTDJ <- RTD[!grepl(regex, transMail)]
    '
 ,@input_data_1 = N'
    SELECT carbrand, transMail
    FROM dbo.mtcarsLQ;'
 ,@input_data_1_name =  N'RT'
 ,@output_data_1_name = N'RTDJ'
WITH RESULT SETS
(
 ("carbrand" NVARCHAR(120) NOT NULL,
  "transMail" NVARCHAR(20) NOT NULL)
);
```

The results are the same two rows shown in Figure 4-8—the rows I found with
the LIKE operator. However, I used a much more sophisticated regular expression,
"^[[:alnum:]._-]+@[[:alnum:].-]+$", which also finds less obvious errors.

Sometimes a combination of the values from two or more variables might be
suspicious. Each value by itself might be plausible; however, the combination might be
weird. Listing 4-12 checks for the average temperature over months in the air quality
dataset.

Listing 4-12. Checking Suspicious Combinations of Values

```
SELECT MONTH(DateM) AS MonthD,
 AVG(Temp) AS tempAvg,
 COUNT(*) AS nRows
FROM dbo.airqualityLQ
GROUP BY MONTH(DateM);
```

You can see the results in Figure 4-9.

	MonthD	tempAvg	nRows
1	2	56	1
2	5	65	31
3	6	79	30
4	7	83	31
5	8	83	31
6	9	76	30
7	10	90	2

Figure 4-9. *Checking the average temperature over months*

Figure 4-9 shows that the average temperature for month 10, October, is higher than the average temperature for any other month, including July and August. This does not look very plausible.

Measuring Data Quality over Time

A side effect of analytical projects is frequently improved data quality. When you analyze the data, you find all kinds of errors and missing values that nobody expected to have in the database. It can be valuable to show the improvements in the data quality to managers, users, and key stakeholders. It is not too complicated to measure data quality over time. All you need to do is create a data quality data warehouse (DQDW). Then, every time you do the data profiling, record the amount of profiled data, the amount of correct data, and the amount of incorrect or suspicious data.

You can use the *star schema* database model for the DQDW, as you use it for any other data warehouse. In a star schema, there are two kinds of tables: *fact tables*, where you store the *measures*, and *dimensions*, which give context to those measures. You use the *attributes* of the dimensions for *aggregating* and breaking down, or pivoting, the measures. There are many books, articles, and presentations about the star schema, but you can easily start with the article at `https://en.wikipedia.org/wiki/Star_schema`.

For DQDW, I suggest at least three dimensions.

- *Date*, which stores the profiling date, with attributes that can help aggregate the data, like months, quarters, and years

- *Table*, where you store the names of the tables you are profiling, including the schema name, database name, server name, application name that created the table, and more

- *Column*, where you store the column names you are profiling, together with the id of the table of the column ID

You can add more dimensions if you can collect the data. For example, if a company has dedicated *data stewards*, employees responsible for specific tables and/or columns and rows, you could add a dimension *Employee*. Listing 4-13 creates a DQDW and the suggested three dimensions.

Listing 4-13. DQDW Dimensions

```
/* Data Quality DW */
USE master;
CREATE DATABASE DQDW;
GO
USE DQDW;
GO
-- Dimensions
-- DimDate
CREATE TABLE dbo.DimDate
(DateId int NOT NULL,
 FullDate DATE NOT NULL);
ALTER TABLE dbo.DimDate
 ADD CONSTRAINT PK_DimDate
  PRIMARY KEY(DateId);
GO
-- DimTable
CREATE TABLE dbo.DimTable
(TableId int NOT NULL,
 TableName sysname NOT NULL,
 SchemaName sysname NOT NULL,
```

```
DatabaseName sysname NULL,
 ServerName sysname NULL,
 ApplicationName sysname NULL);
GO
ALTER TABLE dbo.DimTable
 ADD CONSTRAINT PK_DimTable
  PRIMARY KEY(TableId);
GO
-- DimColumn
CREATE TABLE dbo.DimColumn
(ColumnId int NOT NULL,
 ColumnName sysname NOT NULL,
 TableId int NOT NULL);
GO
ALTER TABLE dbo.DimColumn
 ADD CONSTRAINT PK_DimColumn
  PRIMARY KEY(ColumnId);
GO
ALTER TABLE dbo.DimColumn
 ADD CONSTRAINT FK_DimColumn_DimTable
  FOREIGN KEY(TableId)
  REFERENCES dbo.DimTable(TableId);
GO
```

I suggest the following two fact tables.

- *Tables*, where you measure the table totals, like number of rows, number of rows with NULLs in any column, number of rows with errors, and similar

- Columns, where you measure on the column level

Listing 4-14 creates two fact tables with all the necessary constraints.

Listing 4-14. DQDW Fact Tables

```
-- Fact tables
-- Tables
CREATE TABLE dbo.FactTables
```

```
(TableId int NOT NULL,
 DateId int NOT NULL,
 NumRows bigint NOT NULL,
 NumUnknownRows bigint NOT NULL,
 NumErroneousRows bigint NOT NULL);
GO
ALTER TABLE dbo.FactTables
 ADD CONSTRAINT PK_FactTables
  PRIMARY KEY(TableId, DateId);
GO
ALTER TABLE dbo.FactTables
 ADD CONSTRAINT FK_FactTables_DimTable
  FOREIGN KEY(TableId)
  REFERENCES dbo.DimTable(TableId);
ALTER TABLE dbo.FactTables
 ADD CONSTRAINT FK_FactTables_DimDate
  FOREIGN KEY(DateId)
  REFERENCES dbo.DimDate(DateId);
GO
-- Columns
CREATE TABLE dbo.FactColumns
(ColumnId int NOT NULL,
 DateId int NOT NULL,
 NumValues bigint NOT NULL,
 NumUnknownValues bigint NOT NULL,
 NumErroneousValues bigint NOT NULL);
GO
ALTER TABLE dbo.FactColumns
 ADD CONSTRAINT PK_FactColumns
  PRIMARY KEY(ColumnId, DateId);
GO
ALTER TABLE dbo.FactColumns
 ADD CONSTRAINT FK_FacColumns_DimColumn
  FOREIGN KEY(ColumnId)
  REFERENCES dbo.DimColumn(ColumnId);
```

```
ALTER TABLE dbo.FactColumns
 ADD CONSTRAINT FK_FactColumns_DimDate
  FOREIGN KEY(DateId)
  REFERENCES dbo.DimDate(DateId);
GO
```

The DQDW schema is shown in Figure 4-10. I created the diagram with SQL Server Management Studio.

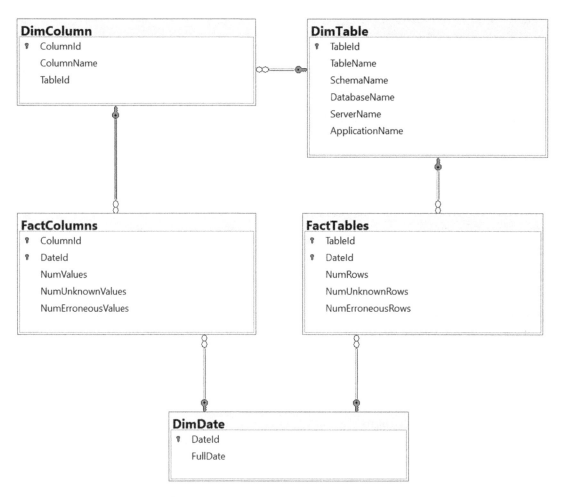

Figure 4-10. *DQDW schema*

Let's return to measuring the data quality. In the next section, I discuss the amount of information in the data.

Measuring the Information

Information is based on *surprise*. When somebody tells you something, you learn something new only if you are surprised. Surprise is similar to *uncertainty*. The more uncertain you are about an outcome, the more times you may be surprised. In data, surprise can be connected to *probability*.

Imagine that you have a box of small balls. There are two colors of balls: yellow and blue. In the box, 80% of the balls are yellow. The probability that you blindly pull a yellow ball from the box is 0.8, and the probability for a blue ball is 0.2. When you pull a ball out of the box, you expect to get a yellow ball, and you are surprised with a blue ball in 20% of cases.

With two possible states (yellow and blue), you are surprised in 50% of cases. When you have an equal number of balls of each color, the information in this box is the maximum. Now let's add a third color, such as red. If the number of balls of each color is still equal, the probability for each color is 0.33. If you still bet on the color when you pull out a single ball, you are surprised in 67% of cases. With four distinct colors and equal distribution of the colors, you would be surprised in 75% of cases. If you have balls of only a single color in the box, you would never be surprised. The information in the box is zero.

Let's recall *equal height discretization*, introduced in Chapter 3. With the balls in the box example, you can see why equal height discretization makes sense. With discretizing, you are lowering the number of distinct states, thus lowering the maximum uncertainty in the variable. With equal height discretization, the uncertainty loss is at the smallest possible.

Introducing Entropy

In information theory, entropy measures the level of information, surprise, or uncertainty. Let's look at the formula for entropy.

$$H(Y) = (-1) * \sum_{i=1}^{n} P(Y_i) * log_2 \big(P(Y_i) \big)$$

The formula works for discrete variables with *n distinct states*. $P(Y_i)$ is the *probability* of the *i-th state*. In the formula, a logarithm with base 2 is used, according to the definition of entropy. However, a logarithm with any base would work the same. The sum of the entropy of all states is the total entropy for the variable. The sum is multiplied by –1 because the probability is between 0 and 1 (greater than 0; otherwise, the state is not in the formula). The logarithmic function returns negative values for this interval. Entropy does not work that well for continuous variables; therefore, I focus on discrete variables only.

Before calculating the actual entropy, let's calculate the maximum entropy for a different number of distinct states. Using the logarithms equations, it is possible to simplify this calculation. Let's look at an example with three states. The probability of each state is 1/3. We can express log(1/3) as the difference between two logarithms.

$$\log\left(\frac{1}{3}\right) = \log(1) - \log(3) = -\log(3)$$

Of course, log(1) is equal to zero. Now let's calculate the total entropy for all three states.

$$(-1)*\left(\left(\frac{1}{3}\right)*(-\log(3)) + \left(\frac{1}{3}\right)*(-\log(3)) + \left(\frac{1}{3}\right)*(-\log(3))\right) = (-1)*\left(\frac{3}{3}\right)*(-\log(3)) = \log(3)$$

Calculating the maximum entropy is simple. Listing 4-15 shows the calculation for a few possibilities.

Listing 4-15. Calculating the Maximum Possible Entropy

```
SELECT LOG(2,2) AS TwoStatesMax,
 LOG(3,2) AS ThreeStatesMax,
 LOG(4,2) AS FourStatesMax,
 LOG(5,2) AS FiveStatesMax;
```

Figure 4-11 shows the result. The maximum entropy grows with the number of distinct states.

	TwoStatesMax	ThreeStatesMax	FourStatesMax	FiveStatesMax
1	1	1.58496250072116	2	2.32192809488736

Figure 4-11. *Maximum entropy for two to five distinct states*

The nice thing about entropy is that it works with only probabilities, not with the variable's values. Therefore, you can use entropy to compare the amount of information between multiple variables, no matter whether they are numbers or strings. You can calculate the actual entropy and the *relative entropy*, which is the actual entropy divided by the maximum entropy. Variables with higher uncertainty are more useful for analyses.

Let's go over calculating the entropy of the hpdescription variable from the mtcars dataset in two steps. The first step calculates the state frequency and the state probability, as Listing 4-16 shows.

Listing 4-16. Calculating the Entropy, Step 1

```
-- Entropy of hpdescription
WITH prob AS
(
SELECT hpdescription,
 COUNT(hpdescription) AS stateFreq
FROM dbo.mtcars
GROUP BY hpdescription
),
stateEntropy AS
(
SELECT hpdescription, stateFreq,
 1.0 * stateFreq / SUM(stateFreq) OVER () AS stateProbability
FROM prob
)
SELECT * FROM stateEntropy;
```

The result is shown in Figure 4-12.

	hpdescription	stateFreq	stateProbability
1	Medium	13	0.406250000000
2	Strong	10	0.312500000000
3	Weak	9	0.281250000000

Figure 4-12. *State frequencies and probabilities*

The second step, shown in Listing 4-17, calculates the entropy.

Listing 4-17. Calculating the Entropy, Step 2

```
-- Entropy of hpdescription
WITH prob AS
(
SELECT hpdescription,
 COUNT(hpdescription) AS stateFreq
FROM dbo.mtcars
GROUP BY hpdescription
),
stateEntropy AS
(
SELECT hpdescription, stateFreq,
 1.0 * stateFreq / SUM(stateFreq) OVER () AS stateProbability
FROM prob
)
--SELECT * FROM stateEntropy;
SELECT 'hpdescription' AS Variable,
 (-1) * SUM(stateProbability * LOG(stateProbability, 2)) AS TotalEntropy,
 LOG(COUNT(*), 2) AS MaxPossibleEntropy,
 100 * ((-1)*SUM(stateProbability * LOG(stateProbability, 2))) /
 (LOG(COUNT(*), 2)) AS PctOfMaxPossibleEntropy
FROM stateEntropy;
```

Figure 4-13 shows the result for the hpdescription variable.

	Variable	TotalEntropy	MaxPossibleEntropy	PctOfMaxPossibleEntropy
1	hpdescription	1.56705242819723	1.58496250072116	98.8700002356033

Figure 4-13. *Entropy of hpdescription*

The same query can calculate the entropy of the cyl variable.

```
-- Entropy of cyl
WITH prob AS
(
SELECT cyl,
 COUNT(cyl) AS stateFreq
```

```
FROM dbo.mtcars
GROUP BY cyl
),
stateEntropy AS
(
SELECT cyl,
 1.0 * stateFreq / SUM(stateFreq) OVER () AS stateProbability
FROM prob
)
--SELECT * FROM stateEntropy
SELECT 'cyl' AS Variable,
 (-1) * SUM(stateProbability * LOG(stateProbability, 2)) AS TotalEntropy,
 LOG(COUNT(*), 2) AS MaxPossibleEntropy,
 100 * ((-1)*SUM(stateProbability * LOG(stateProbability, 2))) /
 (LOG(COUNT(*), 2)) AS PctOfMaxPossibleEntropy
FROM stateEntropy;
```

The total entropy for the `cyl` variable is 1.53099371349313; the relative entropy in percent is 96.5949486373669. The relative entropy of the `hpdescription` variable is slightly higher, meaning that there is a bit more information, or uncertainty, in this variable. Both variables have three distinct states.

Mutual Information

The amount of information in a variable is an important data quality measure. However, with entropy, you can go beyond data profiling. With *mutual information*, you can check for associations between pairs of variables, whether their values are expressed as numbers or strings. Mutual information quantifies the amount of information you can obtain about one variable by knowing some other variable values. Here is the formula for mutual information.

$$I(X;Y) = \sum_{y \in Y}\sum_{x \in X} P(x,y) log_2\left(\frac{P(x,y)}{P(x)P(y)}\right)$$

Listing 4-18 shows the calculation of the mutual information between the
hpdescription and cyl variables. The query is similar to the query to calculate the
entropy.

Listing 4-18. Calculating Mutual Information

```
-- Mutual information I(hpdescription; cyl)
WITH counts AS
(
SELECT cyl AS x,
 hpdescription AS y,
 COUNT(*) AS n
FROM dbo.mtcars
GROUP BY cyl, hpdescription
)
--SELECT * FROM counts
, probs AS
(
SELECT x, y, n,
 1.0 * n / SUM(n) OVER () AS xyProb,
 1.0 * SUM(n) OVER(PARTITION BY x) AS xn,
 1.0 * SUM(n) OVER(PARTITION BY x) / SUM(n) OVER () AS xProb,
 1.0 * SUM(n) OVER(PARTITION BY y) AS yn,
 1.0 * SUM(n) OVER(PARTITION BY y) / SUM(n) OVER () AS yProb
FROM counts
)
--SELECT * FROM probs
--ORDER BY x, y;
SELECT SUM(xyProb * LOG(xyProb / xProb / yProb, 2))
 AS mutualInformation
FROM probs;
```

Figure 4-14 shows the result.

	mutualInformation
1	0.954298454365242

Figure 4-14. *Mutual information between hpdescription and cyl*

It is simple to calculate mutual information between any pair of discrete variables. For example, the following code calculates the carb and engine variables.

```
-- Mutual information carb, engine
WITH counts AS
(
SELECT carb AS x,
 engine AS y,
 COUNT(*) AS n
FROM dbo.mtcars
GROUP BY carb, engine
)
--SELECT * FROM counts
, probs AS
(
SELECT x, y, n,
 1.0 * n / SUM(n) OVER () AS xyProb,
 1.0 * SUM(n) OVER(PARTITION BY x) AS xn,
 1.0 * SUM(n) OVER(PARTITION BY x) / SUM(n) OVER () AS xProb,
 1.0 * SUM(n) OVER(PARTITION BY y) AS yn,
 1.0 * SUM(n) OVER(PARTITION BY y) / SUM(n) OVER () AS yProb
FROM counts
)
--SELECT * FROM probs
--ORDER BY x, y;
SELECT SUM(xyProb * LOG(xyProb / xProb / yProb, 2))
 AS mutualInformation
FROM probs;
```

The result is 0.450596261545634. These two variables are less associated than the previous pair.

Conditional Entropy

Conditional entropy is another measurement. The conditional entropy of the Y variable considering the X variable tells us how much uncertainty is left in Y after knowing X's values. The following is the formula for conditional entropy.

$$H(Y|X) = (-1) * \sum_{y \in Y} \sum_{x \in X} P(x,y) \log_2\left(\frac{P(x,y)}{P(x)}\right)$$

Listing 4-19 calculates the conditional entropy of the `hpdescription` variable considering the `cyl` variable.

Listing 4-19. Conditional Entropy H(hpdescription | cyl)

```
-- Conditional entropy H(hpdescription | cyl)
WITH counts AS
(
SELECT cyl AS x,
 hpdescription AS y,
 COUNT(*) AS n
FROM dbo.mtcars
GROUP BY cyl, hpdescription
)
--SELECT * FROM counts
, probs AS
(
SELECT x, y, n,
 1.0 * n / SUM(n) OVER () AS xyProb,
 1.0 * SUM(n) OVER(PARTITION BY x) AS xn,
 1.0 * SUM(n) OVER(PARTITION BY x) / SUM(n) OVER () AS xProb,
 1.0 * SUM(n) OVER(PARTITION BY y) AS yn,
 1.0 * SUM(n) OVER(PARTITION BY y) / SUM(n) OVER () AS yProb
FROM counts
)
--SELECT * FROM probs
--ORDER BY x, y;
```

```
SELECT (-1) * SUM(xyProb * LOG(xyProb / xProb, 2))
 AS conditionalEntropy
FROM probs;
```

The result is 0.612753460998833. Of course, it is easy to calculate the conditional entropy of cyl considering hpdescription.

```
-- Conditional entropy H(cyl | hpdescription)
WITH counts AS
(
SELECT hpdescription AS x,
 cyl AS y,
 COUNT(*) AS n
FROM dbo.mtcars
GROUP BY hpdescription, cyl
)
--SELECT * FROM counts
, probs AS
(
SELECT x, y, n,
 1.0 * n / SUM(n) OVER () AS xyProb,
 1.0 * SUM(n) OVER(PARTITION BY x) AS xn,
 1.0 * SUM(n) OVER(PARTITION BY x) / SUM(n) OVER () AS xProb,
 1.0 * SUM(n) OVER(PARTITION BY y) AS yn,
 1.0 * SUM(n) OVER(PARTITION BY y) / SUM(n) OVER () AS yProb
FROM counts
)
--SELECT * FROM probs
--ORDER BY x, y;
SELECT (-1) * SUM(xyProb * LOG(xyProb / xProb, 2))
 AS conditionalEntropy
FROM probs;
GO
```

The result is 0.57669474629473. This means that the hpdescription variable better explains the cyl variable than the opposite. This does not mean that we found causality; the difference comes from the fact that the entropy of the hpdescription variable is bigger than the entropy of the `cyl` variable. You can calculate conditional entropy from entropy and mutual information, like the following formula shows.

$$H(Y|X) = H(Y) - I(X;Y)$$

Figure 4-15 shows the relation between entropy, mutual information, and conditional entropy.

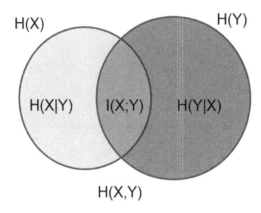

Figure 4-15. *The relation between entropy, mutual information, and conditional entropy*

Now let's check whether the calculation of the conditional entropy was correct. The calculation is shown in Listing 4-20.

Listing 4-20. Validating Conditional Entropy Calculation

```
/*
Variable          TotalEntropy
hpdescription     1.56705242819723
cyl               1.53099371349313

cyl hpdescription mutual information 0.954298454365242
*/
SELECT
 1.56705242819723 - 0.954298454365242 AS '(hpdescription | cyl)',
 1.53099371349313 - 0.954298454365242 AS '(cyl | hpdescription)';
GO
```

Figure 4-16 shows the result.

	(hpdescription | cyl)	(cyl | hpdescription)
1	0.612753973831988	0.576695259127888

Figure 4-16. *Conditional entropies between two variables verified*

Entropy, mutual information, and conditional entropy are very useful measures that go beyond descriptive statistics and make it possible to inspect and analyze numerics and strings at the same time.

Conclusion

In this chapter, you learned about data quality, and specifically about data profiling, which helps you find inaccurate data. When talking about analyses, we usually think about statistics. However, measures from the information theory are very useful.

Finally, I clean my AdventureWorksDW2017 database and my SQL Server instance with the following code.

```
-- Clean up
USE AdventureWorksDW2017;
DROP VIEW IF EXISTS dbo.airqualityLQ;
DROP VIEW IF EXISTS dbo.mtcarsLQ;
DROP EXTERNAL LIBRARY [data.table];
GO
USE master;
DROP DATABASE DQDW;
GO
```

In the next chapter, I discuss issues and queries specific to time-oriented data, data where you have at least one variable that denotes the time.

PART III

Dealing with Time

CHAPTER 5

Time-Oriented Data

Understanding what kind of *temporal* data can appear in a database is very important. Some queries that deal with temporal data are hard to optimize. Data preparation of time series data has many of its own rules.

Queries that deal with time-oriented data are frequent in analytical systems. Besides a simple comparison of data in different periods, much more complex problems and analyses can arise. This chapter and the next chapter deal with time-oriented data. In this chapter, you learn about the following.

- Application and system times

- SQL Server system-version tables and the problems associated with querying them

- Optimization of queries that deal with intervals, especially overlapping intervals

- Smoothing time series data with moving averages

Application and System Times

Dealing with time data means that one or more columns in the table denote the date or the time. A single column marks the date and time of an event. This kind of mark is implemented in practically any business system. For example, every invoice includes the date and likely the due date. However, a simple timestamp dating when something happened or should happen is often not enough.

A *temporal database* uses *time constraints*. Often, we pretend that the state of a database is correct until the end of time. Yet, we know that this is not true in real life. For example, a company is a supplier for another company only when under contract. Practically every food product has a limited shelf time or *period of validity*. "Supplier A is under a contract" is a simple, timeless proposition. Adding the time makes it a

© Dejan Sarka 2021
D. Sarka, *Advanced Analytics with Transact-SQL*, https://doi.org/10.1007/978-1-4842-7173-5_5

timestamped proposition. Time can be limited on one side, such as "Supplier A is under contract from time point A," or on both sides, such as "Supplier A is under contract from time point A to time point B." The words *from* and the *to* denote an interval of validity. You can also express this interval with the word *between*, such as "Supplier A is under contract between time points A and B."

Humans define time validity, known as *application times*. A proposition can also be timestamped with *system times*, which are the times when transactions happen. With system (or *database*) times, you see when a specific state was known to and valid for the database. The database system can automatically maintain these times. SQL Server uses *system-versioned tables* to maintain database times.

Inclusion Constraints

Dealing with temporal databases can be complex. For example, it is hard to optimize queries that search for all rows in a temporal table that was valid in a specific time interval. Searching for overlapping intervals is traditionally a complex issue. But optimizing a query is not temporal queries' biggest problem.

A query must first return the correct result; only after that can you optimize it. The problem arises when you need to join two temporal tables. A simple inner join is not good enough. For example, a supplier can supply a product only in the time intervals it is under contract. But the supplier can be under a contract multiple times in different time intervals and can supply products in any of those intervals. A simple join of the two tables over the supplier ID is not good enough; it would join the rows from incorrect intervals. An *inclusion constraint* is needed. The intervals of the supplied products must be included in one of the supplier's intervals. Table 5-1 is a small example of a timestamped suppliers table.

Table 5-1. *Timestamped Suppliers Table*

SupplierId	From	To
S1	D04	D10
S2	D02	D04
S2	D07	D10

In Table 5-1, the supplier was under contract in two different time intervals. Table 5-2 shows the supplies of the products for the two suppliers in Table 5-1.

Table 5-2.   *Timestamped Product Supplies Table*

SupplierId	ProductId	From	To
S1	P1	D05	D07
S2	P1	D09	D10
S2	P2	D02	D03
S2	P2	D08	D10

If you join the two tables on the supplier ID only, you get rows that should not be in the results. For example, you could get a row with supplier S2 from D02 to D04 joined with the row for supplier S2 product P1 from D09 to D10. This row never existed in the database.

The from and to timepoints in both tables do not give a time unit. The time unit, or the *time granularity*, depends on the business's needs. For contracts, the granularity of a day is usually appropriate. You can always work with integers instead of points of time and use a lookup table to add the units or give the integers time context.

Demo Data

I created the data used in this chapter in the AdventureWorksDW2017 demo database. First, let's create an auxiliary table of numbers and dates. I work on a daily level of granularity. Listing 5-1 creates this auxiliary table.

Listing 5-1.   Creating the Auxiliary Table of Dates and Numbers

```
-- Demo data
USE AdventureWorksDW2017;
GO

-- Auxiliary table of dates and numbers
DROP TABLE IF EXISTS dbo.DateNums;
CREATE TABLE dbo.DateNums
```

```
 (n int NOT NULL PRIMARY KEY,
  d date NOT NULL UNIQUE);
GO

SET NOCOUNT ON;
DECLARE @max AS INT, @rc AS INT, @d AS DATE;
SET @max = 10000;
SET @rc = 1;
SET @d = '20100101'      -- Initial date

INSERT INTO dbo.DateNums VALUES(1, @d);
WHILE @rc * 2 <= @max
BEGIN
  INSERT INTO dbo.DateNums
  SELECT n + @rc, DATEADD(day, n + @rc - 1, @d)
  FROM dbo.DateNums;
  SET @rc = @rc * 2;
END

INSERT INTO dbo.DateNums
  SELECT n + @rc, DATEADD(day, n + @rc - 1, @d)
  FROM dbo.DateNums
  WHERE n + @rc <= @max;
GO

-- Check data
SELECT * FROM dbo.DateNums

SET NOCOUNT OFF;
GO
```

I created a table with 10,000 rows. The dates range from January 1, 2010, to May 18, 2037. Figure 5-1 shows the first few rows of this table.

	n	d
1	1	2010-01-01
2	2	2010-01-02
3	3	2010-01-03
4	4	2010-01-04
5	5	2010-01-05

Figure 5-1. *The content of the dbo.DateNums table*

The next step is to create a demo table of sales, like Listing 5-2 shows. The table represents a fictitious example of sales of packaged fast food, like sandwiches and salads. Every product sold has some validity time. The table has only a few columns; I want to mention four of them. SoldDate and ExpirationDate are the dates when a product was sold and the expiration date. The two integer columns—sn and en—are redundantly denoting the same dates with integers, which I found in the auxiliary table for the dates and the numbers.

Listing 5-2. Creating the Demo Sales Table

```
DROP TABLE IF EXISTS dbo.Sales;
CREATE TABLE dbo.Sales
(
 Id int IDENTITY(1,1) PRIMARY KEY,
 ProductKey int,
 PName varchar(13),
 sn int NULL,
 en int NULL,
 SoldDate date,
 ExpirationDate date
);
GO
```

Listing 5-3 populates the table and checks the data. I inserted the data from the dbo. FactInternetSales and dbo.FactResellerSales tables three times to get more rows. I added a fictitious product name and the CustomerKey column from Internet sales and the ResellerKey column from reseller sales for the fictitious product key. I used the order date for the sell date and added a random expiration date for an interval of validity between one and thirty days.

Listing 5-3. Populating the Demo Sales Table

```
-- Insert the data
DECLARE @y AS int = 0;
WHILE @y < 3
BEGIN
INSERT INTO dbo.Sales
 (ProductKey, PName, SoldDate, ExpirationDate)
SELECT
 CustomerKey AS ProductKey,
 'Product ' + CAST(CustomerKey AS char(5)) AS PName,
 CAST(DATEADD(year, @y *3, OrderDate)
  AS date) AS SoldDate,
 CAST(DATEADD(day,
        CAST(CRYPT_GEN_RANDOM(1) AS int) % 30 + 1,
        DATEADD(year, @y *3, OrderDate))
  AS date) AS ExpirationDate
FROM dbo.FactInternetSales
UNION ALL
SELECT
 ResellerKey AS ProductKey,
 'Product ' + CAST(ResellerKey AS char(5)) AS PName,
 CAST(DATEADD(year, @y *3, OrderDate)
  AS date) AS SoldDate,
 CAST(DATEADD(day,
        CAST(CRYPT_GEN_RANDOM(1) AS int) % 30 + 1,
        DATEADD(year, @y *3, OrderDate))
  AS date) AS ExpirationDate
FROM dbo.FactResellerSales;
SET @y = @y + 1;
END;
GO
```

```
-- Update sn and en
UPDATE dbo.Sales
    SET sn = d.n
FROM dbo.Sales AS s
 INNER JOIN dbo.DateNums AS d
  ON s.SoldDate = d.d;
UPDATE dbo.Sales
    SET en = d.n
FROM dbo.Sales AS s
 INNER JOIN dbo.DateNums AS d
  ON s.ExpirationDate = d.d;
GO

-- Overview
SELECT *
FROM dbo.Sales;
GO
```

Once the insert was finished, I updated the sn and en integer interval columns using the auxiliary date numbers table. Figure 5-2 shows few random rows from this dbo.Sales table.

	Id	ProductKey	PName	sn	en	SoldDate	ExpirationDate
1	311618	299	Product 299	2920	2941	2017-12-29	2018-01-19
2	141395	16409	Product 16409	2323	2327	2016-05-11	2016-05-15
3	301615	22057	Product 22057	3663	3673	2020-01-11	2020-01-21
4	261190	13259	Product 13259	3406	3421	2019-04-29	2019-05-14
5	90800	220	Product 220	1093	1120	2012-12-28	2013-01-24
6	280629	12147	Product 12147	3533	3559	2019-09-03	2019-09-29

Figure 5-2. *The dbo.Sales table content*

Note   I use very short names for auxiliary objects that are in the focus of the queries. In the preceding example, the columns sn and en are the data numbers, represented as dates in the next two date columns.

I use this table later in this chapter when I discuss optimization possibilities for temporal queries.

I want to start with a brief introduction to the system-versioned tables and problems with queries on them.

System-Versioned Tables and Issues

As you probably noticed, I do not spend a lot of time introducing the SQL Server features. I focus on advanced querying. You can learn about the features and language elements in the official SQL Server documentation. Documentation on temporal tables is at `https://docs.microsoft.com/en-us/sql/relational-databases/tables/temporal-tables?view=sql-server-ver15`.

SQL Server system-versioned tables automatically maintain a full history of data changes in a table. The best thing about SQL Server temporal tables is that you can start using them without modifying any existing application. In the next section, you learn how system-versioned tables work.

A Quick Introduction to System-Versioned Tables

Each *temporal table* has an associated *history table*. You can create a history table manually or let SQL Server do it for you automatically. The history of changes can use a *retention period*. You can see only the history up to a specific point in time; for example, it only changes up to one year. Let's start the examples by setting the history retention to true in my AdventureWorksDW2017 database.

```
ALTER DATABASE AdventureWorksDW2017
 SET TEMPORAL_HISTORY_RETENTION ON;
GO
```

In Listing 5-4, I initially created two regular tables. Both tables have an identical structure. The Vf and Vt columns represent the temporal part; the names are short for valid from and valid to. The data type of those columns is *datetime2*. This is a prerequisite for temporal tables.

Listing 5-4. Two Regular Tables with Identical Structure

```
-- T1
DROP TABLE IF EXISTS dbo.T1;
CREATE TABLE dbo.T1
(
 Id INT NOT NULL PRIMARY KEY CLUSTERED,
 C1 INT,
 Vf DATETIME2 NOT NULL,
 Vt DATETIME2 NOT NULL
);
GO
-- T1 hist
DROP TABLE IF EXISTS dbo.T1_Hist;
CREATE TABLE dbo.T1_Hist
(
 Id INT NOT NULL,
 C1 INT,
 Vf DATETIME2 NOT NULL,
 Vt DATETIME2 NOT NULL
);
GO
CREATE CLUSTERED INDEX IX_CL_T1_Hist ON dbo.T1_Hist(Vt, Vf);
GO
```

Listing 5-5 populates the two tables with demo rows and shows the content of the tables. Please note that they are still regular tables.

Listing 5-5. Populating the Temporal Tables

```
-- Populate tables
INSERT INTO dbo.T1_Hist(Id, C1, Vf, Vt) VALUES
(1,1,'20191101','20191106'),
(1,2,'20191106','20210202');
GO
INSERT INTO dbo.T1(Id, C1, Vf, Vt) VALUES
(1,3,'20210202','99991231 23:59:59.9999999');
```

```
GO
SELECT *
FROM dbo.T1;
SELECT *
FROM dbo.T1_Hist;
GO
```

Figure 5-3 shows the content of the tables.

	Id	C1	Vf	Vt
1	1	3	2021-02-02 00:00:00.0000000	9999-12-31 23:59:59.9999999

	Id	C1	Vf	Vt
1	1	1	2019-11-01 00:00:00.0000000	2019-11-06 00:00:00.0000000
2	1	2	2019-11-06 00:00:00.0000000	2021-02-02 00:00:00.0000000

Figure 5-3. *Initial content of the two temporal tables*

Now let's convert the two tables to system-versioned ones. I use the ALTER TABLE dbo.T1 command twice. First, I connect the two datetime2 columns to a period. Then I connect both tables to the system version. I define the history table and the retention period—three months in this case, as shown in Listing 5-6.

Listing 5-6. Converting Tables to Temporal Ones

```
-- Convert to temporal
ALTER TABLE dbo.T1 ADD PERIOD FOR SYSTEM_TIME (Vf, Vt);
GO
ALTER TABLE dbo.T1 SET
(
SYSTEM_VERSIONING = ON
 (
  HISTORY_TABLE = dbo.T1_Hist,
  HISTORY_RETENTION_PERIOD = 3 MONTHS
 )
);
GO
```

Listing 5-7 does an insert, an update, and a delete in the *current* table. Then both tables are queried.

Listing 5-7. Updating Temporal Tables

```
-- Do some updates
INSERT INTO dbo.T1 (Id, C1)
VALUES (2, 1), (3, 1);

WAITFOR DELAY '00:00:01';

UPDATE dbo.T1
   SET C1 = 2
 WHERE Id = 2;

WAITFOR DELAY '00:00:01';

DELETE FROM dbo.T1
WHERE id = 3;
GO
-- Show the content of the two tables
SELECT *
FROM dbo.T1;
SELECT *
FROM dbo.T1_Hist;
GO
```

Figure 5-4 shows the result.

	Id	C1	Vf	Vt
1	1	3	2021-02-02 00:00:00.0000000	9999-12-31 23:59:59.9999999
2	2	2	2021-02-05 10:30:10.6341329	9999-12-31 23:59:59.9999999

	Id	C1	Vf	Vt
1	1	1	2019-11-01 00:00:00.0000000	2019-11-06 00:00:00.0000000
2	1	2	2019-11-06 00:00:00.0000000	2021-02-02 00:00:00.0000000
3	2	1	2021-02-05 10:30:10.6184612	2021-02-05 10:30:10.6341329
4	3	1	2021-02-05 10:30:10.6184612	2021-02-05 10:30:10.6341329

Figure 5-4. *The content of the temporal tables after the updates*

Before updating or deleting a row, the old version is inserted automatically in the history table. In addition, the Vt (valid to) column is updated. The current rows have the maximum datetime2 value in the Vt column.

An application refers to the current table only. The application does not need to know that there is an underlying history table in the database. However, the application could use poorly written queries, as in Listing 5-7, using the SELECT * clause instead of listing the columns explicitly. Such an application would suddenly get two new datetime2 columns from a query if you added these two columns after the application was already in production. You can use temporal tables, even if the application is written poorly. You can hide the two period columns, as shown in Listing 5-8.

Listing 5-8. Hiding the Period Columns

```
-- Hiding the period columns
ALTER TABLE dbo.T1 ALTER COLUMN Vf ADD HIDDEN;
ALTER TABLE dbo.T1 ALTER COLUMN Vt ADD HIDDEN;
GO
-- Show the content of the two tables
SELECT *
FROM dbo.T1;
SELECT *
FROM dbo.T1_Hist;
GO
```

Figure 5-5 shows the result of the queries. The two period columns are hidden in the result set of the current table. You could still explicitly request them in a query.

	Id	C1
1	1	3
2	2	2

	Id	C1	Vf	Vt
1	1	1	2019-11-01 00:00:00.0000000	2019-11-06 00:00:00.0000000
2	1	2	2019-11-06 00:00:00.0000000	2021-02-02 00:00:00.0000000
3	2	1	2021-02-05 10:30:10.6184612	2021-02-05 10:30:10.6341329
4	3	1	2021-02-05 10:30:10.6184612	2021-02-05 10:30:10.6341329

Figure 5-5. *Result of SELECT * with hidden period columns*

You can query both the current and history tables in a single query with the FOR SYSTEM_TIME clause in the FROM part of a query. This clause has the following options.

- AS OF: Shows the state of the table on a specific time point

- FROM timepoint TO timepoint: All rows with all versions that were active during the interval. Rows that were active on interval boundaries are not included.

- BETWEEN timepoint AND timepoint: All rows with all versions that were active during the interval. Rows that were active on the upper interval boundary are included.

- CONTAINED IN (timepoint, timepoint): All rows were opened and closed in the period, including upper and lower boundaries. This clause gives you an idea of what was changed in a specified period.

- ALL: Returns the union of the current and history table rows.

The following is an example of the AS OF query.

```
-- AS OF
SELECT Id, C1, Vf, Vt
FROM dbo.T1
 FOR SYSTEM_TIME
  AS OF '2021-02-05 10:30:10.6184612';
```

For brevity, I do not show the result of this query. Listing 5-9 shows how the FROM..TO and BETWEEN clauses work.

Note   I retrieved the last valid from date (the Vf column) from the dbo.T1_Hist table in a variable, because when you run the code, your last updated date differs from the one I had when I wrote this chapter. If you uncomment the SELECT @Vf code, you can see the date you are using.

Listing 5-9. The FROM..TO and BETWEEN Clauses

```
-- FROM..TO and BETWEEN
DECLARE @Vf AS DATETIME2;
SET @Vf =
 (SELECT MAX(Vf) FROM dbo.T1_Hist);
-- SELECT @Vf;SELECT Id, C1, Vf, Vt
FROM dbo.T1
 FOR SYSTEM_TIME
  FROM '2019-11-06 00:00:00.0000000'
    TO '2021-02-05 10:30:10.6184612';
SELECT Id, C1, Vf, Vt
FROM dbo.T1
 FOR SYSTEM_TIME
  BETWEEN '2019-11-06 00:00:00.0000000'
      AND '2021-02-05 10:30:10.6184612';
```

Note that the code uses the same interval boundaries for both queries. However, the result of the queries is slightly different, as Figure 5-6 shows.

	Id	C1	Vf	Vt
1	1	3	2021-02-02 00:00:00.0000000	9999-12-31 23:59:59.9999999
2	1	2	2019-11-06 00:00:00.0000000	2021-02-02 00:00:00.0000000

	Id	C1	Vf	Vt
1	1	3	2021-02-02 00:00:00.0000000	9999-12-31 23:59:59.9999999
2	1	2	2019-11-06 00:00:00.0000000	2021-02-02 00:00:00.0000000
3	2	1	2021-02-05 10:30:10.6184612	2021-02-05 10:30:10.6341329
4	3	1	2021-02-05 10:30:10.6184612	2021-02-05 10:30:10.6341329

Figure 5-6. *Different results of the FROM..TO and BETWEEN clauses*

Listing 5-10 shows the usage of the CONTAINED IN clause.

Listing 5-10. The CONTAINED IN Clause

```
-- CONTAINED IN
DECLARE @Vf AS DATETIME2;
SET @Vf =
 (SELECT MAX(Vf) FROM dbo.T1_Hist);
SET @Vf = DATEADD(s, 3, @Vf);
-- SELECT @Vf;
SELECT Id, C1, Vf, Vt
FROM dbo.T1
 FOR SYSTEM_TIME
  CONTAINED IN
   ('2019-11-06 00:00:00.0000000',
    '2021-02-06 00:00:00.0000000');
```

Figure 5-7 shows the result, which are all changes in the defined period.

	Id	C1	Vf	Vt
1	1	2	2019-11-06 00:00:00.0000000	2021-02-02 00:00:00.0000000
2	2	1	2021-02-05 10:30:10.6184612	2021-02-05 10:30:10.6341329
3	3	1	2021-02-05 10:30:10.6184612	2021-02-05 10:30:10.6341329

Figure 5-7. *All changes in a period*

I focus on the ALL clause in the next section because you can get unexpected results with this clause.

Querying System-Versioned Tables Surprises

So far, the implementation of the system-versioned tables looked great. Microsoft solved only the smaller problem, the system time, and not the application time, which you must still solve by yourself. However, the same unpleasant surprises are waiting for you when you query the temporal system tables with the FOR SYSTEM_TIME ALL clause.

Let's start by querying the dbo.T1 table. If you remember the results from Figure 5-4, there are two current and four historical rows in this table (and the associated history table). Let's execute the query in Listing 5-11 and include the actual execution plan.

Listing 5-11. Using the FOR SYSTEM_TIME ALL Clause with Retention Period

```
SELECT Id, C1, Vf, Vt
FROM dbo.T1
 FOR SYSTEM_TIME ALL;
```

Figure 5-8 shows the result and the important part of the execution plan.

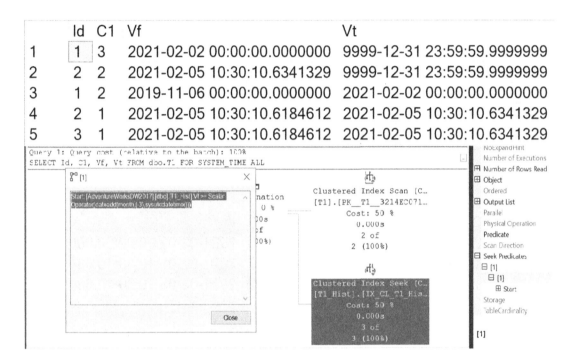

	Id	C1	Vf	Vt
1	1	3	2021-02-02 00:00:00.0000000	9999-12-31 23:59:59.9999999
2	2	2	2021-02-05 10:30:10.6341329	9999-12-31 23:59:59.9999999
3	1	2	2019-11-06 00:00:00.0000000	2021-02-02 00:00:00.0000000
4	2	1	2021-02-05 10:30:10.6184612	2021-02-05 10:30:10.6341329
5	3	1	2021-02-05 10:30:10.6184612	2021-02-05 10:30:10.6341329

Figure 5-8. *Rows older than the retention period are filtered out*

Note   You should execute this code on the same day you create and populate the tables; otherwise, you might get a different result. You might get fewer rows because the retention period could have already passed.

In the result, you can see that there are only five rows. The oldest row from the history table was automatically filtered out. SQL Server added the following seek predicate to the query.

```
Seek Keys[1]: Start: [AdventureWorksDW2017].[dbo].[T1_Hist].Vt >= Scalar
Operator(dateadd(month,(-3),sysutcdatetime()))
```

Remember that I altered the AdventureWorksDW2017 database to set the retention period to three months? Now you see how this retention period is implemented. The old rows are still in the history table. They filtered automatically. If you want to clean them, you must delete them manually. However, to delete rows from the history table, you must break the temporal connection between the current and the history table, delete the rows, and establish a connection between the two tables again. This is a clumsy process. To avoid further surprises, let's remove the retention period and drop the two demo tables.

```
-- Clean up
ALTER TABLE dbo.T1 SET (SYSTEM_VERSIONING = OFF);
DROP TABLE IF EXISTS dbo.T1;
DROP TABLE IF EXISTS dbo.T1_Hist;
GO
ALTER DATABASE AdventureWorksDW2017
 SET TEMPORAL_HISTORY_RETENTION OFF;
GO
```

Another surprise when querying temporal tables relates to the granularity of the beginning and ending time point of the validity period. You need to use the *datetime2* data type. You can specify the number of digits for this data type, from 0 to 7, with 7 as the default. Zero digits mean a granularity of 1 second; seven digits mean a granularity of 100 nanoseconds. Listing 5-12 creates a temporal table with one-second granularity and inserts the first row.

Listing 5-12. Temporal Table with One-Second Granularity

```
-- Granularity issues - granularity 1s demo
CREATE TABLE dbo.T1
(
 id INT NOT NULL CONSTRAINT PK_T1 PRIMARY KEY,
 c1 INT NOT NULL,
 vf DATETIME2(0) GENERATED ALWAYS AS ROW START NOT NULL,
 vt DATETIME2(0) GENERATED ALWAYS AS ROW END NOT NULL,
 PERIOD FOR SYSTEM_TIME (vf, vt)
)
```

```
WITH (SYSTEM_VERSIONING = ON (HISTORY_TABLE = dbo.T1_Hist));
GO
-- Initial row
INSERT INTO dbo.T1(id, c1)
VALUES(1, 1);
GO
```

Now let's execute two transactions in a single batch, as shown in Listing 5-13, and then immediately query the table with the FOR SYSTEM_TIME ALL clause. Please note that the delay between the two updates is one-tenth of a second, which is less than the granularity specified when I created the table.

Listing 5-13. Two Transactions in a Short Time

```
BEGIN TRAN
UPDATE dbo.T1
   SET c1 = 2
 WHERE id=1;
COMMIT;
WAITFOR DELAY '00:00:00.1';
BEGIN TRAN
UPDATE dbo.T1
   SET c1 = 3
 WHERE id=1;
COMMIT;
GO
SELECT *
FROM dbo.T1
FOR SYSTEM_TIME ALL;
GO
```

Figure 5-9 shows that there are only two rows in the result. I also included the execution plan and highlighted the following predicate.

```
[AdventureWorksDW2017].[dbo].[T1_Hist].[vf]<>[AdventureWorksDW2017].[dbo].
[T1_Hist].[vt]
```

Again, this predicate was added automatically.

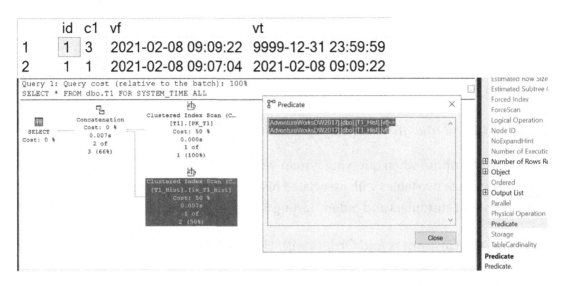

	id	c1	vf	vt
1	1	3	2021-02-08 09:09:22	9999-12-31 23:59:59
2	1	1	2021-02-08 09:07:04	2021-02-08 09:09:22

Figure 5-9. *Filtering the rows where the starting time of an interval is equal to the ending time*

If you query both tables manually, you see a "missing" row in the result. I query the tables with the following code, which cleans up my database after I no longer need the two tables.

```
SELECT *
FROM dbo.T1
UNION ALL
SELECT *
FROM dbo.T1_Hist;
GO
-- Clean up
ALTER TABLE dbo.T1 SET (SYSTEM_VERSIONING = OFF);
DROP TABLE IF EXISTS dbo.T1;
DROP TABLE IF EXISTS dbo.T1_Hist;
GO
```

Figure 5-10 shows the result. You see that the "missing" row is there. In the third row, the "valid from" time is equal to the "valid to" time because the granularity is one second.

	id	c1	vf	vt
1	1	3	2021-02-08 09:09:22	9999-12-31 23:59:59
2	1	1	2021-02-08 09:07:04	2021-02-08 09:09:22
3	1	2	2021-02-08 09:09:22	2021-02-08 09:09:22

Figure 5-10. All rows from the current and history tables

The third surprise when querying system-versioned tables is, in my opinion, the worst. Let's create two tables with associated history tables in the tempdb database. The tables represent customers and orders. Listing 5-14 shows this code.

Listing 5-14. Customers and Orders with History Tables

```
-- Missing Inclusion Constraints
USE tempdb;
GO
-- Customers
CREATE TABLE dbo.CustomerHistory
(
 CId CHAR(3) NOT NULL,
 CName CHAR(5) NOT NULL,
 ValidFrom DATETIME2 NOT NULL,
 ValidTo DATETIME2 NOT NULL
);
CREATE CLUSTERED INDEX CIX_CustomerHistory ON dbo.CustomerHistory
(ValidTo ASC, ValidFrom ASC);
GO
CREATE TABLE dbo.Customer
(
 CId CHAR(3) PRIMARY KEY,
 CName CHAR(5) NOT NULL
);
GO
-- Orders
CREATE TABLE dbo.OrdersHistory
(
```

```
 OId CHAR(3) NOT NULL,
 CId CHAR(3) NOT NULL,
 Q INT NOT NULL,
 ValidFrom DATETIME2 NOT NULL,
 ValidTo DATETIME2 NOT NULL
);
CREATE CLUSTERED INDEX CIX_OrdersHistory ON dbo.OrdersHistory
(ValidTo ASC, ValidFrom ASC);
GO
CREATE TABLE dbo.Orders
(
 OId CHAR(3) NOT NULL PRIMARY KEY,
 CId CHAR(3) NOT NULL,
 Q INT NOT NULL
);
GO
ALTER TABLE dbo.Orders ADD CONSTRAINT FK_Orders_Customer
 FOREIGN KEY (CId) REFERENCES dbo.Customer(CID);
GO
```

Now let's fill these tables with demo data, as Listing 5-15 shows.

Listing 5-15. Inserting Demo Data

```
-- Demo data
INSERT INTO dbo.CustomerHistory
 (CId, CName, ValidFrom, ValidTo)
VALUES
 ('111','AAA','20180101', '20180201'),
 ('111','BBB','20180201', '20180220');
INSERT INTO dbo.Customer
 (CId, CName)
VALUES
 ('111','CCC');
INSERT INTO dbo.OrdersHistory
 (OId, CId, Q, ValidFrom, ValidTo)
```

```
VALUES
 ('001','111',1000,'20180110','20180201'),
 ('001','111',2000,'20180201','20180203'),
 ('001','111',3000,'20180203','20180220');
INSERT INTO dbo.Orders
 (OId, CId, q)
VALUES
 ('001','111',4000);
GO
-- Check the data
SELECT * FROM dbo.Customer;
SELECT * FROM dbo.CustomerHistory;
SELECT * FROM dbo.Orders;
SELECT * FROM dbo.OrdersHistory;
GO
```

Figure 5-11 shows the demo data in all four tables.

	CId	CName		
1	111	CCC		

	CId	CName	ValidFrom	ValidTo
1	111	AAA	2018-01-01 00:00:00.0000000	2018-02-01 00:00:00.0000000
2	111	BBB	2018-02-01 00:00:00.0000000	2018-02-20 00:00:00.0000000

	OId	CId	Q	
1	001	111	4000	

	OId	CId	Q	ValidFrom	ValidTo
1	001	111	1000	2018-01-10 00:00:00.0000000	2018-02-01 00:00:00.0000000
2	001	111	2000	2018-02-01 00:00:00.0000000	2018-02-03 00:00:00.0000000
3	001	111	3000	2018-02-03 00:00:00.0000000	2018-02-20 00:00:00.0000000

Figure 5-11. *Demo data*

Please note the timeline of the data. I have one customer and one order.

1. The initial state, on January 1, 2018, was a customer named AAA; no orders.

2. On January 10, 2018, a single order was inserted, with a quantity of 1,000.

3. On February 2, 2018, the customer's name was updated to BBB, and the order quantity was raised to 2,000.

4. On February 3, 2018, the order quantity was raised to 3,000.

5. On February 20, 2018, the customer's name was updated to CCC, and the order quantity was raised to 4,000.

Listing 5-16 makes the tables temporal and shows all the data in both tables.

Listing 5-16. Making the Tables Temporal

```
-- Make the tables temporal
-- Alter the current tables
ALTER TABLE dbo.Customer
 ADD ValidFrom DATETIME2 GENERATED ALWAYS AS ROW START NOT NULL
  CONSTRAINT DFC_StartDate1 DEFAULT '20180220 00:00:00.0000000',
 ValidTo DATETIME2 GENERATED ALWAYS AS ROW END NOT NULL
  CONSTRAINT DFC_EndDate1 DEFAULT '99991231 23:59:59.9999999',
 PERIOD FOR SYSTEM_TIME (ValidFrom, ValidTo);
GO
ALTER TABLE dbo.Orders
 ADD ValidFrom DATETIME2 GENERATED ALWAYS AS ROW START NOT NULL
  CONSTRAINT DFO_StartDate1 DEFAULT '20180220 00:00:00.0000000',
 ValidTo DATETIME2 GENERATED ALWAYS AS ROW END NOT NULL
  CONSTRAINT DFO_EndDate1 DEFAULT '99991231 23:59:59.9999999',
 PERIOD FOR SYSTEM_TIME (ValidFrom, ValidTo);
GO
-- Enable system versioning
ALTER TABLE dbo.Customer
 SET (SYSTEM_VERSIONING = ON
  (HISTORY_TABLE = dbo.CustomerHistory,
   DATA_CONSISTENCY_CHECK = ON));
GO
ALTER TABLE dbo.Orders
 SET (SYSTEM_VERSIONING = ON
  (HISTORY_TABLE = dbo.OrdersHistory,
   DATA_CONSISTENCY_CHECK = ON));
```

```
GO
-- Check all data
SELECT *
FROM dbo.Customer
 FOR SYSTEM_TIME ALL;
SELECT *
FROM dbo.Orders
 FOR SYSTEM_TIME ALL;
GO
```

Figure 5-12 shows the content of the tables. Of course, these are the same rows in Listing 5-11, unioned to two rowsets instead of four.

	CId	CName	ValidFrom	ValidTo
1	111	CCC	2018-02-20 00:00:00.0000000	9999-12-31 23:59:59.9999999
2	111	AAA	2018-01-01 00:00:00.0000000	2018-02-01 00:00:00.0000000
3	111	BBB	2018-02-01 00:00:00.0000000	2018-02-20 00:00:00.0000000

	OId	CId	Q	ValidFrom	ValidTo
1	001	111	4000	2018-02-20 00:00:00.0000000	9999-12-31 23:59:59.9999999
2	001	111	1000	2018-01-10 00:00:00.0000000	2018-02-01 00:00:00.0000000
3	001	111	2000	2018-02-01 00:00:00.0000000	2018-02-03 00:00:00.0000000
4	001	111	3000	2018-02-03 00:00:00.0000000	2018-02-20 00:00:00.0000000

Figure 5-12. *The content of the temporal tables*

Imagine that there is a view in the database that joined the customers and orders tables before the tables were made temporal. Listing 5-17 shows the code for the view.

Listing 5-17. A View That Joins Customers and Orders

```
CREATE OR ALTER VIEW dbo.CustomerOrders
AS
SELECT C.CId, C.CName, O.OId, O.Q,
 c.ValidFrom AS CVF, c.ValidTo AS CVT,
 o.ValidFrom AS OVF, o.ValidTo AS OVT
```

```
FROM dbo.Customer AS c
 INNER JOIN dbo.Orders AS o
  ON c.CId = o.CId;
GO
```

Let's query the view. The queries in Listing 5-18 show the state on specific time points.

Listing 5-18. Querying the View on Specific Time Points

```
-- Current state
SELECT *
FROM dbo.CustomerOrders;
-- Specific dates - propagated to source tables
SELECT *
FROM dbo.CustomerOrders
 FOR SYSTEM_TIME AS OF '20180102';
SELECT *
FROM dbo.CustomerOrders
 FOR SYSTEM_TIME AS OF '20180131';
SELECT *
FROM dbo.CustomerOrders
 FOR SYSTEM_TIME AS OF '20180201';
SELECT *
FROM dbo.CustomerOrders
 FOR SYSTEM_TIME AS OF '20180203';
SELECT *
FROM dbo.CustomerOrders
 FOR SYSTEM_TIME AS OF '20180221';
GO
```

Figure 5-13 shows the results.

	CId	CName	Old	Q	CVF	CVT
1	111	CCC	001	4000	2018-02-20 00:00:00.0000000	9999-12-31 23:59:59.9999999

	CId	CName	Old	Q	CVF	CVT	OVF	OVT

	CId	CName	Old	Q	CVF	CVT
1	111	AAA	001	1000	2018-01-01 00:00:00.0000000	2018-02-01 00:00:00.0000000

	CId	CName	Old	Q	CVF	CVT
1	111	BBB	001	2000	2018-02-01 00:00:00.0000000	2018-02-20 00:00:00.0000000

	CId	CName	Old	Q	CVF	CVT
1	111	BBB	001	3000	2018-02-01 00:00:00.0000000	2018-02-20 00:00:00.0000000

	CId	CName	Old	Q	CVF	CVT
1	111	CCC	001	4000	2018-02-20 00:00:00.0000000	9999-12-31 23:59:59.9999999

Figure 5-13. *The state on the specific time points through a view*

Now comes the surprise. Listing 5-19 queries the view by using the FOR SYSTEM_TIME ALL clause.

Listing 5-19. Querying the View with the FOR SYSTEM_TIME ALL Clause

```
SELECT *
FROM dbo.CustomerOrders
 FOR SYSTEM_TIME ALL;
```

Figure 5-14 shows the result.

	CId	CName	OId	Q	CVF	CVT
1	111	CCC	001	4000	2018-02-20 00:00:00.0000000	9999-12-31 23:59:59.9999999
2	111	CCC	001	1000	2018-02-20 00:00:00.0000000	9999-12-31 23:59:59.9999999
3	111	CCC	001	2000	2018-02-20 00:00:00.0000000	9999-12-31 23:59:59.9999999
4	111	CCC	001	3000	2018-02-20 00:00:00.0000000	9999-12-31 23:59:59.9999999
5	111	AAA	001	4000	2018-01-01 00:00:00.0000000	2018-02-01 00:00:00.0000000
6	111	AAA	001	1000	2018-01-01 00:00:00.0000000	2018-02-01 00:00:00.0000000
7	111	AAA	001	2000	2018-01-01 00:00:00.0000000	2018-02-01 00:00:00.0000000
8	111	AAA	001	3000	2018-01-01 00:00:00.0000000	2018-02-01 00:00:00.0000000
9	111	BBB	001	4000	2018-02-01 00:00:00.0000000	2018-02-20 00:00:00.0000000
10	111	BBB	001	1000	2018-02-01 00:00:00.0000000	2018-02-20 00:00:00.0000000
11	111	BBB	001	2000	2018-02-01 00:00:00.0000000	2018-02-20 00:00:00.0000000
12	111	BBB	001	3000	2018-02-01 00:00:00.0000000	2018-02-20 00:00:00.0000000

Figure 5-14. *Spurious rows in the result set*

For me, this was a surprise. There are all combinations of all states for the customer and the order in the result set. Some rows show the state that never existed. For example, there was never a part of the database the combination where the customer's name is CCC and order quantity is 1,000. What went wrong? Let's analyze the execution plan, the plan shown in Figure 5-15.

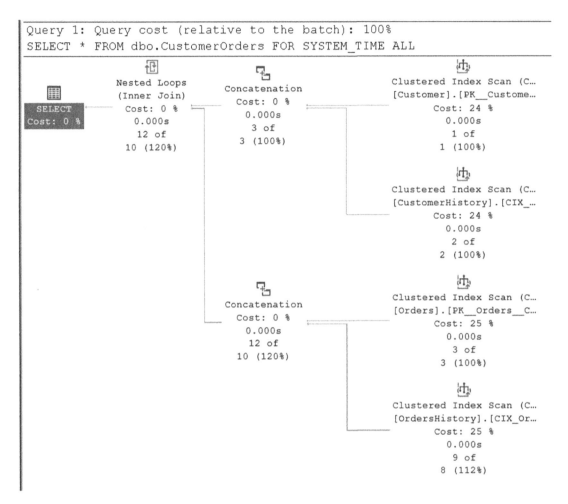

Query 1: Query cost (relative to the batch): 100%
SELECT * FROM dbo.CustomerOrders FOR SYSTEM_TIME ALL

Figure 5-15. Execution plan of a query on the view

The plan is very naïve. SQL Server unions customers with their history, orders with their history, and then joins them with regular nested loops. What is missing here is an inclusion constraint. If you remember, I explained the inclusion constraints at the beginning of this chapter. I warned that without them, you could get an incorrect result when you join two rowsets.

This is a big issue for me. Now, whenever you use the FOR SYSTEM_TIME ALL clause, you must know exactly what you are querying—a source table or a view that maybe joins two or more temporal tables. This is opposite from one of the main principles of the relational model. That principle says that querying base or derived tables (views) should always be the same, that an application should not know whether it queries a table or a view.

The following code cleans the `tempdb` database.

```
-- Clean up
DROP VIEW IF EXISTS dbo.CustomerOrders;
ALTER TABLE dbo.Orders SET (SYSTEM_VERSIONING = OFF);
ALTER TABLE dbo.Orders DROP PERIOD FOR SYSTEM_TIME;
DROP TABLE IF EXISTS dbo.Orders;
DROP TABLE IF EXISTS dbo.OrdersHistory;
ALTER TABLE dbo.Customer SET (SYSTEM_VERSIONING = OFF);
ALTER TABLE dbo.Customer DROP PERIOD FOR SYSTEM_TIME;
DROP TABLE IF EXISTS dbo.Customer;
DROP TABLE IF EXISTS dbo.CustomerHistory;
GO
```

Now let's switch to another problem you might get when querying temporal tables—the performance of the queries that search for overlapping intervals.

Optimizing Temporal Queries

Let's switch back to the AdventureWorksDW2017 demo database to the `dbo.Sales` table, which has two columns that define the application validity time. Which rows in this table were valid at least once during a given interval? This means searching for the rows where the interval of validity *overlaps* with the given interval. Figure 5-16 shows a few rows from this table.

	Id	ProductKey	PName	sn	en	SoldDate	ExpirationDate
1	311618	299	Product 299	2920	2941	2017-12-29	2018-01-19
2	141395	16409	Product 16409	2323	2327	2016-05-11	2016-05-15
3	301615	22057	Product 22057	3663	3673	2020-01-11	2020-01-21
4	261190	13259	Product 13259	3406	3421	2019-04-29	2019-05-14
5	90800	220	Product 220	1093	1120	2012-12-28	2013-01-24
6	280629	12147	Product 12147	3533	3559	2019-09-03	2019-09-29

Figure 5-16. *The content of the dbo.Sales table*

Please note that Figure 5-15 is the same as Figure 5-2. The validity interval is denoted in two ways: the two date columns, SoldDate and ExpiredDate, and the two integer columns, sn and en.

Two intervals, i_1 and i_2, overlap if the beginning of the first interval, b_1, is lower or equal to the end of the second interval, e_2, and the beginning of the second interval, b_2, is lower than or equal to the end of the first interval e1, as the following formula shows.

$$\left(i_1 \text{ overlaps } i_2\right) \Leftrightarrow \left(b_1 \leq e_2\right) \wedge \left(b_2 \leq e_1\right)$$

Before querying the table, I created the optimal indexes for the overlapping queries with the following code.

```
USE AdventureWorksDW2017;
GO
-- Create indexes on the interval boundaries
CREATE INDEX idx_sn
 ON dbo.Sales(sn) INCLUDE(en);
CREATE INDEX idx_en
 ON dbo.Sales(en) INCLUDE(sn);
GO
```

I used the integer validity interval. The first index is on the starting point and includes the ending point of the interval, and the second on the ending point and includes the starting point. Listing 5-20 seeks overlapping intervals when the given interval is somewhere close to the beginning of the timeline for the data and when the given interval is close to the end of it. I set the statistics to see the number of logical reads.

Listing 5-20. Searching for Overlapping Intervals at the Beginning and End

```
SET STATISTICS IO ON;
GO
-- beginning of data
DECLARE
 @sn AS INT = 370,
 @en AS INT = 400;
SELECT Id, sn, en
FROM dbo.Sales
```

```
WHERE sn <= @en AND en >= @sn
OPTION (RECOMPILE);      -- preventing plan reusage
GO
-- index seeks idx_sn
-- logical reads: 6

-- end of data
DECLARE
 @sn AS INT = 3640,
 @en AS INT = 3670;
SELECT Id, sn, en
FROM dbo.Sales
WHERE sn <= @en AND en >= @sn
OPTION (RECOMPILE);      -- preventing plan reusage
GO
-- index seeks idx_en
-- logical reads: 21
```

From the comments in the code, you can see that 6 logical reads were needed for the first query and 21 for the second. Both queries used an efficient execution plan. The first query did the index seek using the index where the key is at the beginning of the interval. The second query did the index seek using the index where the key is at the end of the interval.

Note You can get a slightly different number of logical reads due to many factors, such as index fragmentation. In addition, you can get a slightly different number of rows because when the table was populated (see Listing 5-3), the expiration date was set randomly.

Now let's search for the overlapping intervals when the searched interval is around the middle of the timeline, as shown in Listing 5-21.

Listing 5-21. Searching for Overlapping Intervals in the Middle

```
-- middle of data
DECLARE
 @sn AS INT = 1780,
 @en AS INT = 1810;
SELECT Id, sn, en
FROM dbo.Sales
WHERE sn <= @en AND en >= @sn
OPTION (RECOMPILE);      -- preventing plan reuse
GO
-- index seeks idx_sn
-- logical reads: 299
```

The query used 299 logical reads to return 1,609 rows. Figure 5-17 shows the execution plan.

```
Query 1: Query cost (relative to the batch): 100%
SELECT Id, sn, en FROM dbo.Sales WHERE sn <= @en AND en >= @sn OPTION (RECOMPILE)
```

Index Seek (NonClustere...
[Sales].[idx_sn]
Cost: 100 %
0.009s
1609 of
105517 (1%)

SELECT
Cost: 0 %

Figure 5-17. *Execution plan for the query in Listing 5-21*

The query used the index where the key is the starting point of the intervals. The index covered a query. Still, there were a lot of logical reads. What happened is that the query could quickly find with an index seek the starting point of the first interval that does not overlap with the searched interval. In this interval, the beginning is later than the start of the searched interval. However, from here on, the query needed to do a partial scan of the covering index. The query scanned approximately half of the pages of the index. This is a known problem.

One frequently proposed solution is to use the special *relational interval tree* indexes. You can find an example of this solution in T-SQL in "Interval Queries in SQL Server," *an* article by Itzik Ben-Gan at `https://www.itprotoday.com/sql-server/interval-queries-sql-server`. I do not repeat the solution here. The solution works; however, it is way too complex. In addition, the performance improvements, although substantial, are still not overly impressive.

I show two other solutions here. Before explaining them, note that there is no single solution that would work in any case or for any kind of data. Every solution has its advantages and drawbacks.

Modifying the Filter Predicate

I like my first solution because it is extremely simple. All you need to do is to slightly modify the filter predicate in the query's `WHERE` clause. Let's explain the idea. In the query in Listing 5-21, the seek excluded the rows from the end of the timeline. Is it possible to exclude the rows at the beginning of the timeline using the same index, where the key is at the beginning of the intervals?

The answer to the previous question is yes. However, it is not possible to exclude all the non-overlapping rows from the right side. The idea is to exclude the rows where the start of the interval is early enough to make it impossible for the interval to overlap with the searched interval. Figure 5-18 shows the idea.

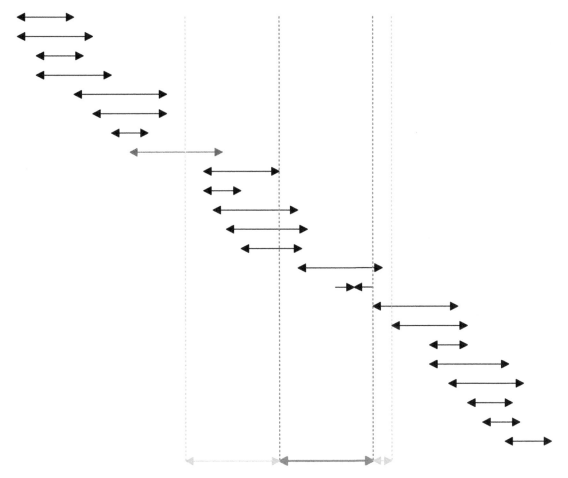

Figure 5-18. *The idea for modifying the filter predicate*

In the figure, I use the black color for showing the intervals in my table. The blue color marks the searched interval (at the bottom of the figure). The starting points define the order of the intervals. Eliminating intervals from the right side of the searched interval is done by eliminating all intervals where the beginning point is later than the ending point of the searched interval. The boundary in the figure is marked with a dotted vertical yellow line on the right side of the searched interval.

Eliminating from the left side is more complex. You must go to the left, away from the beginning of the searched interval, for at least for the length of the longest interval in the table. The longest interval is marked with red color. The intervals that begin earlier than the left dotted vertical yellow line cannot overlap with the searched interval.

After the filter on the left side is implemented, some intervals in the result set do not overlap with the searched interval. The filter returns some *false positives*. You get rid of them with a partial scan. This partial scan is scanning fewer pages than the query with the unmodified filter predicate does. Listing 5-22 shows the query with the modified predicate. I know the length of the longest interval since I created the data. The length is 30.

Listing 5-22. Searching for Overlapping Intervals in the Middle with a Modified Filter Predicate

```
-- middle of data
-- Max length of an interval is 30
DECLARE
 @sn AS INT = 1780,
 @en AS INT = 1810;
SELECT Id, sn, en
FROM dbo.Sales
WHERE sn <= @en AND sn >= @sn - 30
  AND en >= @sn AND en <= @en + 30
OPTION (RECOMPILE);     -- preventing plan reusage
GO
-- index seeks idx_sn
-- logical reads: 9
```

The query returned the same 1,609 rows; this time, with only nine logical reads. What I like about this solution is that it is extremely simple. You slightly the filter predicate of the query, and you get improved performance. However, the solution has drawbacks. First, you need to know the length of the longest interval. This is not very complicated to learn. But the longer the longest interval is, the less efficient the query is because it filters out fewer rows from one side. For example, the length of the longest interval could be 900. Look at the query in Listing 5-23.

Listing 5-23. Using a Very Long Longest Interval

```
-- middle of data
-- Max length of an interval is 900
DECLARE
 @sn AS INT = 1780,
 @en AS INT = 1810;
```

171

```
SELECT Id, sn, en
FROM dbo.Sales
WHERE sn <= @en AND sn >= @sn - 900
  AND en >= @sn AND en <= @en + 900
OPTION (RECOMPILE);       -- preventing plan reusage
GO
-- index seeks idx_sn
-- logical reads: 250
```

The query needed 250 logical reads for the same 1,609 rows. It is still better than the original query. With the modified search predicate, in the worst case, this query's performance can be the same as that of the original query.

Using the Unpacked Form

The next solution uses the *unpacked* form of the original table. The unpacked form means that there is a row for every time unit of the interval in the original row. Imagine that you originally have a row with an interval (5, 8) in it. You create four rows with intervals (5, 5), (6, 6), (7, 7), (8, 8). You can create the unpacked form with a join to the auxiliary date numbers table created at the beginning of this chapter.

I created a view based on the query and indexed the view so my SQL Server updates the data automatically when the source data is updated, as shown in Listing 5-24.

Listing 5-24. Creating the Unpacked Form with an Indexed View

```
-- Unpacked form with an indexed view
-- Create view Intervals_Unpacked
DROP VIEW IF EXISTS dbo.SalesU;
GO
CREATE VIEW dbo.SalesU
WITH SCHEMABINDING
AS
SELECT i.id, n.n
FROM dbo.Sales AS i
 INNER JOIN dbo.DateNums AS n
  ON n.n BETWEEN i.sn AND i.en;
GO
```

172

```
-- Index the view
CREATE UNIQUE CLUSTERED INDEX PK_SalesU
 ON dbo.SalesU(n, id);
GO
```

Now let's query the indexed view. This means completely modifying the original query. Listing 5-25 shows the query that uses the unpacked form through the indexed view and searches for overlapping intervals in the middle of the timeline.

Listing 5-25. Querying the Unpacked Form

```
-- Overlapping interval - middle of data
DECLARE
 @sn AS INT = 1780,
 @en AS INT = 1810;
SELECT id
FROM dbo.SalesU
WHERE n BETWEEN @sn AND @en
GROUP BY id
ORDER BY id
OPTION (RECOMPILE);       -- preventing plan reusage
GO
-- index seek in the clustered view
-- logical reads: 43
```

The query used 43 logical reads only to return the same 1,609 rows. This is much more efficient than the original query. However, this query used other resources much more intensively, like Figure 5-19 shows.

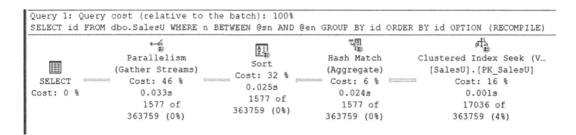

Figure 5-19. *The execution plan of the query on the unpacked form*

SQL Server executed the query in multiple parallel threads. The two black arrows in a yellow circle on the operators in the plan show that these operations were executed in parallel. Therefore, the query used more CPU than other queries I have used so far. This is one of the drawbacks of the solution with the unpacked form. Another drawback is the unpacked form itself—it occupies additional space in the database. Listing 5-26 measures the space.

Listing 5-26. Measuring the Space Used of the Unpacked Form

```
-- Space used
EXEC sys.sp_spaceused 'dbo.Sales';
EXEC sys.sp_spaceused 'dbo.SalesU';
GO
```

Figure 5-20 shows the result.

	name	rows	reserved	data	index_size	unused
1	Sales	363759	31896 KB	17328 KB	13232 KB	1336 KB

	name	rows	reserved	data	index_size	unused
1	SalesU	5838903	98440 KB	98136 KB	224 KB	80 KB

Figure 5-20. *Space used by the unpacked form*

Note   Again, your results might vary slightly.

The space used in the unpacked form is a couple of times bigger than the original table occupies. Therefore, the unpacked form looks less attractive than modifying the search predicate solution. Still, the solution is much simpler than using the relational interval trees.

Time Series

The unpacked form consists of intervals that are single time points. When you have ordered time points, you have a *time series*. You can use time-series data for analyses over time and for forecasting. Time series data can be nicely presented with *line charts*.

In the AdventureWorksDW2017 demo database, the time-series data was already prepared in the dbo.vTimeSeries view. The following code gives you a quick overview of the data in the view.

```
SELECT TOP 5 *
FROM dbo.vTimeSeries;
```

Figure 5-21 shows the result.

	ModelRegion	TimeIndex	Quantity	Amount	CalendarYear	Month	ReportingDate
1	M200 Europe	201012	1	3399.99	2010	12	2010-12-25 00:00:00.000
2	M200 Europe	201101	5	16924.95	2011	1	2011-01-25 00:00:00.000
3	M200 Europe	201102	6	20349.94	2011	2	2011-02-25 00:00:00.000
4	M200 Europe	201103	5	16949.95	2011	3	2011-03-25 00:00:00.000
5	M200 Europe	201104	5	16949.95	2011	4	2011-04-25 00:00:00.000

Figure 5-21. *Time series data*

The view aggregates sales data for each model in each sales region on a monthly level. I aggregate the entire monthly sales for three years to get 36 rows in my time-series data, as shown in Listing 5-27.

Listing 5-27. Preparing the Time Series

```
SELECT TimeIndex,
 SUM(Quantity*2) - 200 AS Quantity,
 DATEFROMPARTS(TimeIndex / 100, TimeIndex % 100, 1) AS DateIndex
FROM dbo.vTimeSeries
WHERE TimeIndex > 201012    -- December 2010 outlier
GROUP BY TimeIndex
ORDER BY TimeIndex;
```

I used a small calculation on the quantity to make the differences between consecutive months bigger. Figure 5-22 shows the partial data.

	TimeIndex	Quantity	DateIndex
1	201101	88	2011-01-01
2	201102	88	2011-02-01
3	201103	100	2011-03-01
4	201104	114	2011-04-01
5	201105	148	2011-05-01

Figure 5-22. *Aggregated time series data*

Now let's start working on the time series.

Moving Averages

If the time series data is changing too much between two consecutive time points, it is difficult to spot the trend and do the forecasts. You can smooth the data by using the *moving averages* instead of the original values. There are many ways to calculate moving averages. Let's start with *simple moving averages*, which is the average of the last few values. The following formula shows the last three values.

$$SMA_i = \left(\sum_{i-2}^{i} v_i \right) / 3$$

It is easy to calculate simple moving averages in T-SQL with window aggregate functions, as Listing 5-28 shows.

Listing 5-28. Simple Moving Averages

```
-- Simple - last 3 values
WITH TSAggCTE AS
(
SELECT TimeIndex,
 SUM(Quantity*2) - 200 AS Quantity,
 DATEFROMPARTS(TimeIndex / 100, TimeIndex % 100, 1) AS DateIndex
FROM dbo.vTimeSeries
WHERE TimeIndex > 201012    -- December 2010 outlier, too small value
GROUP BY TimeIndex
)
```

```
SELECT TimeIndex,
 Quantity,
 AVG(Quantity)
 OVER (ORDER BY TimeIndex
       ROWS BETWEEN 2 PRECEDING
       AND CURRENT ROW) AS SMA,
 DateIndex
FROM TSAggCTE
ORDER BY TimeIndex;
GO
```

Figure 5-23 shows the partial result.

	TimeIndex	Quantity	SMA	DateIndex
1	201101	88	88	2011-01-01
2	201102	88	88	2011-02-01
3	201103	100	92	2011-03-01
4	201104	114	100	2011-04-01
5	201105	148	120	2011-05-01
6	201106	260	174	2011-06-01
7	201107	176	194	2011-07-01
8	201108	186	207	2011-08-01

Figure 5-23. *Simple moving averages*

Weighted moving averages are preferred to simple moving averages, especially in financial applications. When calculating weighted moving averages, you add weight to each value you use in the calculation. Typically, you give a bigger weight to the latest values and a lower to the older values, thus making later values more important. Here is the formula for the weighted moving averages that use the last two values.

$$WMA_i = \left(\sum_{i-1}^{i} v_i * w_i \right) / 2$$

$$\sum_{1}^{2} w_i = 1$$

Note that the sum of all weights is 1. Listing 5-29 calculates the weighted moving averages, using the last two values, with the weight for the latest value 0.7 and the previous values 0.3.

Listing 5-29. Calculating the Weighted Moving Average

```
DECLARE  @A AS FLOAT;
SET @A = 0.7;
WITH TSAggCTE AS
(
SELECT TimeIndex,
 SUM(Quantity*2) - 200 AS Quantity,
 DATEFROMPARTS(TimeIndex / 100, TimeIndex % 100, 1) AS DateIndex
FROM dbo.vTimeSeries
WHERE TimeIndex > 201012    -- December 2010 outlier, too small value
GROUP BY TimeIndex
)
SELECT TimeIndex,
 Quantity,
 AVG(Quantity)
 OVER (ORDER BY TimeIndex
       ROWS BETWEEN 2 PRECEDING
       AND CURRENT ROW) AS SMA,
 @A * Quantity + (1 - @A) *
  ISNULL((LAG(Quantity)
          OVER (ORDER BY TimeIndex)), Quantity)  AS WMA,
 DateIndex
FROM TSAggCTE
ORDER BY TimeIndex;
GO
```

Figure 5-24 compares the simple moving average and the weighted moving average.

	TimeIndex	Quantity	SMA	WMA	DateIndex
1	201101	88	88	88	2011-01-01
2	201102	88	88	88	2011-02-01
3	201103	100	92	96.4	2011-03-01
4	201104	114	100	109.8	2011-04-01
5	201105	148	120	137.8	2011-05-01
6	201106	260	174	226.4	2011-06-01
7	201107	176	194	201.2	2011-07-01
8	201108	186	207	183	2011-08-01

Figure 5-24. *Weighted moving averages*

I used Power BI to show the original values, the simple and the weighted moving averages in a single line chart, shown in Figure 5-25.

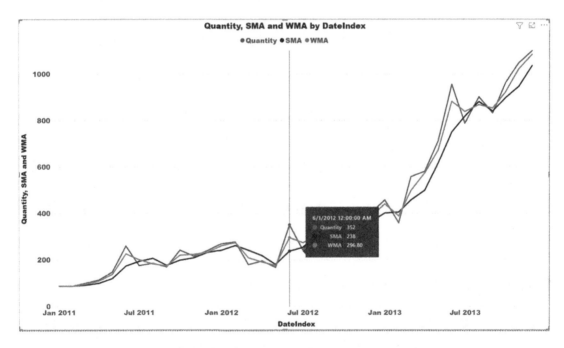

Figure 5-25. *Comparing original values with moving averages*

The line for the simple moving averages is the smoothest, while the line for the original values has the highest peaks and the lowest valleys.

Smoothing is useful for forecasting because the high peaks and the low valleys in the data have a lower influence on the trend. In the line chart visualization over a time series in Power BI, you can add forecasts. Figure 5-26 shows two line charts side by side, one for the original values and one for the simple moving averages, with the forecasts for the next six time points, for the next six months after the last month in the data.

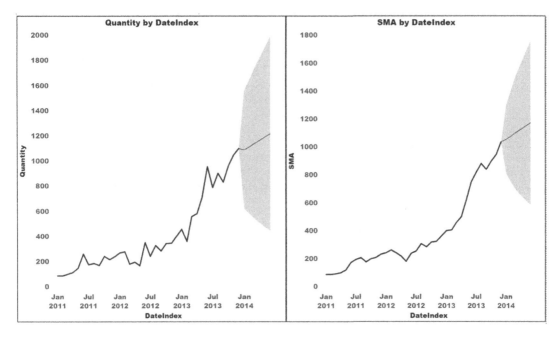

Figure 5-26. *Forecasting with Power BI*

Besides the forecasted values, the graphs show the 95% confidence interval. When I used the moving averages instead of the original values, the confidence interval is narrower.

Conclusion

This chapter is the first part of dealing with time-oriented data. You learned about the application and the system times. This chapter warned you about potential issues with system-versioned table queries. You learned how to optimize queries that search for overlapping intervals. Finally, you learned about time series and moving averages.

It is time to clean my AdventureWorksDW2017 database with the following code.

```
-- Clean up
USE AdventureWorksDW2017;
DROP VIEW IF EXISTS dbo.SalesU;
DROP TABLE IF EXISTS dbo.Sales;
ALTER TABLE dbo.T1 SET (SYSTEM_VERSIONING = OFF);
DROP TABLE IF EXISTS dbo.T1;
DROP TABLE IF EXISTS dbo.T1_Hist;
GO
ALTER DATABASE AdventureWorksDW2017
 SET TEMPORAL_HISTORY_RETENTION ON;
GO
```

Figure 5-26 shows forecasting in a line chart visualization by Power BI. Can you forecast in T-SQL? I show you how to do it in the next chapter, together with more interesting business analyses over time.

CHAPTER 6

Time-Oriented Analyses

Who are my top customers, and what are my top-selling products? How long is a customer faithful to the supplier or the subscribed services and service provider? Which are the most likely days to lose a customer? What are the sales for the next few periods? This chapter shows how to answer these questions using T-SQL.

In Chapter 5, you learned the basics of time-oriented data and some advanced querying problems with this data. By the end of that chapter, you knew about moving averages. This chapter continues with another version of moving averages that is very useful for forecasting. Then it switches to other important business analysis topics. You learn about the following.

- Exponential moving average and forecasting

- ABC analysis

- Survival and hazard analysis

Demo Data

Let's start this chapter with data preparation. For the exponential moving average, I use the same time series data as in the previous chapter. For other analyses, I use the sales data from the AdventureWorksDW2017 database. Let's start with an overview of annual total sales by unioning the Internet and the reseller sales, as shown in Listing 6-1.

Listing 6-1. Checking the AdventureWorks Sales

```
USE AdventureWorksDW2017;
GO
-- Finding orders distribution by year of the order date
WITH od AS
```

© Dejan Sarka 2021
D. Sarka, *Advanced Analytics with Transact-SQL*, https://doi.org/10.1007/978-1-4842-7173-5_6

```
(
SELECT OrderDate
FROM dbo.FactInternetSales
UNION ALL
SELECT OrderDate
FROM dbo.FactResellerSales
)
SELECT YEAR(OrderDate) AS OrderYear,
  MIN(OrderDate) AS minO,
  MAX(OrderDate) AS maxO,
  COUNT(*) AS NumOrders
FROM od
GROUP BY YEAR(OrderDate)
ORDER BY OrderYear;
```

Figure 6-1 shows the result. You can see that most of the sales happened in the years 2011 to 2013.

	OrderYear	minO	maxO	NumOrders
1	2010	2010-12-29 00:00:00.000	2010-12-31 00:00:00.000	366
2	2011	2011-01-01 00:00:00.000	2011-12-31 00:00:00.000	12046
3	2012	2012-01-01 00:00:00.000	2012-12-31 00:00:00.000	25530
4	2013	2013-01-01 00:00:00.000	2013-12-31 00:00:00.000	81341
5	2014	2014-01-01 00:00:00.000	2014-01-28 00:00:00.000	1970

Figure 6-1. *AdventureWorks sales by year*

Listing 6-2 prepares two views that filter out the years 2010 and 2014 from Internet and reseller sales. Then yearly sales are checked again.

Listing 6-2. Preparing the Views with Three Years of Sales Data Only

```
DROP VIEW IF EXISTS dbo.vInternetSales;
GO
CREATE VIEW dbo.vInternetSales
AS
SELECT *
FROM dbo.FactInternetSales
```

```
WHERE YEAR(OrderDate) BETWEEN 2011 AND 2013;
GO
DROP VIEW IF EXISTS dbo.vResellersales;
GO
CREATE VIEW dbo.vResellerSales
AS
SELECT *
FROM dbo.FactResellerSales
WHERE YEAR(OrderDate) BETWEEN 2011 AND 2013;
GO
-- Check the data
WITH od AS
(
SELECT OrderDate
FROM dbo.vInternetSales
UNION ALL
SELECT OrderDate
FROM dbo.vResellerSales
)
SELECT YEAR(OrderDate) AS OrderYear,
  MIN(OrderDate) AS minO,
  MAX(OrderDate) AS maxO,
  COUNT(*) AS NumOrders
FROM od
GROUP BY YEAR(OrderDate)
ORDER BY OrderYear;
GO
```

I do not show the result from this code; the result is the sales for the years from 2011 to 2013.

Exponential Moving Average

The *exponential moving average* is another weighted moving average that is even more popular for financial analyses than the weighted moving average introduced in Chapter 5. The reason for the popularity is because this moving average included calculating all previous values, from the beginning of the data you have, but also gives more importance to the latest values. You see this immediately from the formula.

$$EMA_i = a * v_i + (1-a) * EMA_{i-1}$$

In the calculation, the last value is multiplied by the weight a, and then the previous exponential moving average is added, multiplied by the weight $(1 - a)$. The weight a is a decimal number between zero and one. Suppose you expand the formula and express the exponential moving average in the previous time point using the same formula. In that case, you are using in the calculation the value from the previous time point. You can continue this expansion until you realize that you are using all the previous values to calculate the current moving average; however, the importance of the values from the past drops exponentially. The first exponential moving average is equal to one, or the first value in the time series.

The formula for the exponential moving average does not look very suitable for a set-oriented calculation. It looks much more like a row-oriented problem. You store the moving average in a variable, move to the next row, and calculate the next moving average. Indeed, it is simple to calculate the exponential moving average (EMA) with a cursor, as shown in Listing 6-3.

Listing 6-3. Calculating EMA with a Cursor

```
DECLARE @CurrentEMA AS FLOAT, @PreviousEMA AS FLOAT,
 @t AS INT, @q AS FLOAT,
 @A AS FLOAT;
DECLARE @EMA AS TABLE(TimeIndex INT, Quantity FLOAT, EMA FLOAT);
SET @A = 0.7;

DECLARE EMACursor CURSOR FOR
WITH TSAggCTE AS
(
SELECT TimeIndex,
```

```
  SUM(1.0*Quantity*2) - 200 AS Quantity,
  DATEFROMPARTS(TimeIndex / 100, TimeIndex % 100, 1) AS DateIndex
FROM dbo.vTimeSeries
WHERE TimeIndex > 201012      -- December 2010 outlier, too small value
GROUP BY TimeIndex
)
SELECT TimeIndex, Quantity
FROM TSAggCTE
ORDER BY TimeIndex;

OPEN EMACursor;

FETCH NEXT FROM EMACursor
  INTO @t, @q;
SET @CurrentEMA = @q;
SET @PreviousEMA = @CurrentEMA;

WHILE @@FETCH_STATUS = 0
BEGIN
  SET @CurrentEMA = @A*@q + (1-@A)*@PreviousEMA;
  INSERT INTO @EMA (TimeIndex, Quantity, EMA)
   VALUES(@t, @q, @CurrentEMA);
   SET @PreviousEMA = @CurrentEMA;
  FETCH NEXT FROM EMACursor
   INTO @t, @q;
END;

CLOSE EMACursor;
DEALLOCATE EMACursor;

SELECT TimeIndex, Quantity, EMA
FROM @EMA;
GO
```

Figure 6-2 shows the result. I did not bother to round the numbers.

	TimeIndex	Quantity	EMA
1	201101	88	88
2	201102	88	88
3	201103	100	96.4
4	201104	114	108.72
5	201105	148	136.216
6	201106	260	222.8648
7	201107	176	190.05944
8	201108	186	187.217832

Figure 6-2. *EMA calculated with a cursor*

I bet you know that using cursors in T-SQL is not very efficient; however, it is possible to calculate the EMA with a set-oriented query.

Calculating EMA Efficiently

Let's calculate the EMA step by step, starting with the first time series. For simplicity, EMA is denoted with a single letter e, and the weight $(1 - a)$ is denoted with the letter b.

$$e_1 = v_1$$

$$e_2 = a*v_2 + b*e_1 = a*v_2 + b*v_1$$

$$e_3 = a*v_3 + b*(a*v_2 + b*v_1) == a*v_3 + ab*v_2 + b^2*v_1$$

Let's introduce the expanded formula for the fourth EMA.

$$e_4 = a*v_4 + ab*v_3 + ab^2*v_2 + b^3*v_1$$

Let's now divide the whole right side with b on the fourth degree.

$$e_4 = b^4 * \left(\frac{a}{b^4}*v_4 + \frac{a}{b^3}*v_3 + \frac{a}{b^2}*v_2 + \frac{1}{b}*v_4 \right)$$

Now it is possible to calculate the EMA for every single row with the help of the window aggregate functions. Let's first calculate the exponent for the weight $b(1 - a)$ with the query in Listing 6-4.

Listing 6-4. Calculating the Exponent for the EMA Weights

```
DECLARE @A AS FLOAT = 0.7, @B AS FLOAT;
SET @B = 1 - @A;
WITH TSAggCTE AS
(
SELECT TimeIndex,
 SUM(1.0*Quantity*2) - 200 AS Quantity,
 DATEFROMPARTS(TimeIndex / 100, TimeIndex % 100, 1) AS DateIndex
FROM dbo.vTimeSeries
WHERE TimeIndex > 201012    -- December 2010 outlier
GROUP BY TimeIndex
),
EMACTE AS
(
SELECT TimeIndex, Quantity,
  ROW_NUMBER() OVER (ORDER BY TimeIndex) - 1 AS Exponent
FROM TSAggCTE
)
SELECT TimeIndex, Quantity, Exponent
FROM EMACTE;
```

Figure 6-3 shows the partial result.

	TimeIndex	Quantity	Exponent
1	201101	88.0	0
2	201102	88.0	1
3	201103	100.0	2
4	201104	114.0	3
5	201105	148.0	4
6	201106	260.0	5

Figure 6-3. *The exponent for the EMA for every single row*

Listing 6-5 is the full calculation for the complete time series.

Listing 6-5. Calculating EMA with a Set-Oriented Query

```
DECLARE @A AS FLOAT = 0.7;
WITH TSAggCTE AS
(
SELECT TimeIndex,
 SUM(1.0*Quantity*2) - 200 AS Quantity,
 DATEFROMPARTS(TimeIndex / 100, TimeIndex % 100, 1) AS DateIndex
FROM dbo.vTimeSeries
WHERE TimeIndex > 201012     -- December 2010 outlier
GROUP BY TimeIndex
),
EMACTE AS
(
SELECT TimeIndex, Quantity,
  ROW_NUMBER() OVER (ORDER BY TimeIndex) - 1 AS Exponent
FROM TSAggCTE
)
SELECT TimeIndex, Quantity,
 ROUND(
  SUM(CASE WHEN Exponent = 0 THEN 1
           ELSE @A
      END
      * POWER((1 - @A), -Exponent)
      * Quantity
  )
  OVER (ORDER BY TimeIndex)
  * POWER((1 - @A), Exponent)
 , 2) AS EMA
FROM EMACTE;
GO
```

Figure 6-4 is the result of the set-oriented query; this time with the rounding for a more readable result set.

	TimeIndex	Quantity	EMA
1	201101	88.0	88
2	201102	88.0	88
3	201103	100.0	96.4
4	201104	114.0	108.72
5	201105	148.0	136.22
6	201106	260.0	222.86
7	201107	176.0	190.06
8	201108	186.0	187.22

Figure 6-4. *EMA calculated and rounded*

The EMA is very frequently used for forecasting. This is a simple process, as you see in the next section.

Forecasting with EMA

When you forecast with EMA, you declare the value in the next time point, the first time point in the future, as the EMA in the last time point where you still have the actual value. Then you can move to two time points in the future and calculate the next EMA as the next forecasted value. You can continue this process recursively or in a loop. This again does not sound very efficient because it is a loop and not a set-oriented query. However, you forecast only the next few values with the EMA. You do not want to use the EMA for the forecasts in the distant future. You understand why very soon. Let's first store the results of the EMA calculation in a table with the following code.

```
DROP TABLE IF EXISTS dbo.EMA;
DECLARE @A AS FLOAT = 0.7;
WITH TSAggCTE AS
(
SELECT TimeIndex,
 CAST(SUM(1.0*Quantity*2) - 200 AS FLOAT) AS Quantity,
 DATEFROMPARTS(TimeIndex / 100, TimeIndex % 100, 1) AS DateIndex
FROM dbo.vTimeSeries
WHERE TimeIndex > 201012
GROUP BY TimeIndex
),
```

```
EMACTE AS
(
SELECT TimeIndex, Quantity,
  ROW_NUMBER() OVER (ORDER BY TimeIndex) - 1 AS Exponent
FROM TSAggCTE
)
SELECT
 ROW_NUMBER() OVER(ORDER BY TimeIndex) AS TimeRN,
 TimeIndex, Quantity,
 ROUND(
  SUM(CASE WHEN Exponent = 0 THEN 1
           ELSE @A
      END
      * POWER((1 - @A), -Exponent)
      * Quantity
  )
  OVER (ORDER BY TimeIndex)
  * POWER((1 - @A), Exponent)
 , 2) AS EMA
INTO dbo.EMA
FROM EMACTE;
GO
```

Now let's do the forecasting for the six time points. Listing 6-6 forecasts the sales for the first six months of 2014. The code inserts new rows with forecasted values into the table.

Listing 6-6. Forecasting with EMA

```
-- Selecting the last row
DECLARE @t AS INT,  @v AS FLOAT, @e AS FLOAT;
DECLARE @A AS FLOAT = 0.7, @r AS INT = 36;
SELECT @t = TimeIndex, @v = Quantity, @e = EMA
FROM dbo.EMA
WHERE TimeIndex =
      (SELECT MAX(TimeIndex) FROM dbo.EMA);
-- SELECT @t, @v, @e;
```

```
-- First forecast time point
SET @t = 201401;
SET @r = 37;
-- Forecasting in a loop
WHILE @t <= 201406
BEGIN
 SET @v = ROUND(@A * @v + (1 - @A) * @e, 2);
 SET @e = ROUND(@A * @v + (1 - @A) * @e, 2);
 INSERT INTO dbo.EMA
 SELECT @r, @t, @v, @e;
 SET @t += 1;
 SET @r += 1;
END
SELECT TimeRN, TimeIndex, Quantity, EMA
FROM dbo.EMA
ORDER BY TimeIndex;
GO
```

Figure 6-5 shows the last few rows of the result. Remember, the last six rows show the forecasted values.

	TimeRN	TimeIndex	Quantity	EMA
31	31	201307	788	812.31
32	32	201308	902	875.09
33	33	201309	832	844.93
34	34	201310	962	926.88
35	35	201311	1048	1011.66
36	36	201312	1100	1073.5
37	37	201401	1092.05	1086.49
38	38	201402	1090.38	1089.21
39	39	201403	1090.03	1089.78
40	40	201404	1089.95	1089.9
41	41	201405	1089.94	1089.93
42	42	201406	1089.94	1089.94

Figure 6-5. *Sales forecasted with the EMA*

The forecasted values quickly limit to a constant value (close to 1,090 in the example). Figure 6-6 shows this graphically.

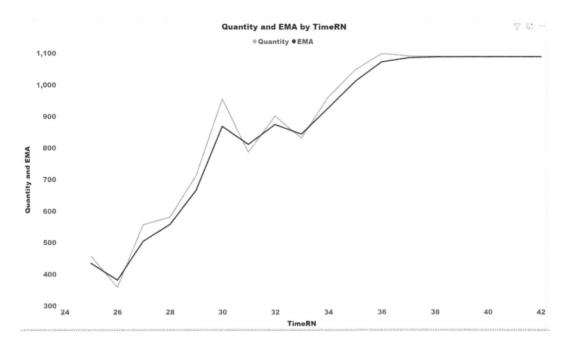

Figure 6-6. *EMA forecasts*

It is easy to understand what is happening here. EMA is smoothing the values, and the same EMA is then used for forecasts. Of course, the differences between the forecasted values become lower and lower the more you go into the future until you are left with a constant value. Therefore, the EMA is useful for very short-term forecasting only. For long-term forecasting, you need something that can show you the trend. In Chapter 7, I show you how to use the linear regression algorithm for calculating the trend.

ABC Analysis

This section deals with analyses that are not strictly related to time. Of course, you can always add the time component in the queries. Nevertheless, the analyses described here are common and useful. The first part describes a lesser-known problem called *relational division*. Under the scientific name hides a less complex question, such as which customers buy all your product categories and which only part of them.

Finding the most important customers and products is one of the most frequent analyses, either over time or without the time. This analysis is known as an *ABC analysis* and the *Pareto principle*.

Relational Division

Let's start with a formal definition of relational division. A *divisor* relation partitions a *dividend* relation to produce a *quotient* relation. The dividend and the divisor relations are in a many-to-many relationship. The quotient relation is made up of those values of one attribute from the dividend relation, for which the second column of the intermediate relation (the resolution of the many-to-many relationship) contains all the values from the divisor relation. Figure 6-7 offers a clearer idea of how to do relational division in T-SQL.

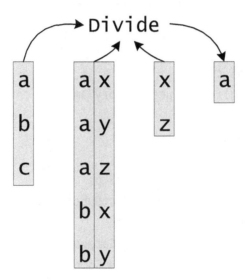

Figure 6-7. *Relational division*

The leftmost relation {a, b, c} is the dividend. The third relation from the left, the relation {x, z} is the divisor. The relation between them, the second relation from the left, is the resolution of the many-to-many relationship between the dividend and the divisor. Only the element {a} of the dividend is related to every single element {x, y} of the divisor. The rightmost relation {a} is the quotient.

In real life, the relational division provides the answer to questions like "the customers subscribed to all services" or "which sales department employees sold each product at least once." Usually, there is also the time component in these questions, like "the sales department employees who sold each product at least once in the current year."

In the AdventureWorksDW2017 demo database, there are four distinct product categories, bikes, clothing, accessories, and components. You can check them with the following query.

```
SELECT ProductCategoryKey, EnglishProductCategoryName
FROM dbo.DimProductCategory;
```

The components are not selling. Therefore, let's search only for the customers who purchased from the other three categories. Listing 6-7 returns the customer key and the number of distinct product categories they purchased from.

Listing 6-7. Listing Customers with the Number of Distinct Categories

```
SELECT s.CustomerKey AS cid,
  COUNT(DISTINCT pc.EnglishProductCategoryName) AS cnt
FROM dbo.FactInternetSales AS s
 INNER JOIN dbo.DimProduct AS p
  ON p.ProductKey = s.ProductKey
 INNER JOIN dbo.DimProductSubcategory AS ps
  ON ps.ProductSubcategoryKey = p.ProductSubcategoryKey
 INNER JOIN dbo.DimProductCategory AS pc
  ON pc.ProductCategoryKey = ps.ProductCategoryKey
WHERE pc.EnglishProductCategoryName <> N'Components'
GROUP BY s.CustomerKey
ORDER BY s.CustomerKey;
```

For brevity, the results are not shown. It is simple to develop the query that lists only the customers who purchased all product categories. You just need to add a filter condition after the aggregation in the HAVING clause. You are searching for the customers who purchased many distinct categories in the data without components. Listing 6-8 shows this query, which is the solution for relational division.

Listing 6-8. Relational Division

```
SELECT s.CustomerKey AS cid,
  MIN(c.LastName + N', ' + c.FirstName) AS cnm
FROM dbo.FactInternetSales AS s
 INNER JOIN dbo.DimProduct AS p
  ON p.ProductKey = s.ProductKey
 INNER JOIN dbo.DimProductSubcategory AS ps
  ON ps.ProductSubcategoryKey = p.ProductSubcategoryKey
 INNER JOIN dbo.DimProductCategory AS pc
  ON pc.ProductCategoryKey = ps.ProductCategoryKey
 INNER JOIN dbo.DimCustomer AS c
  ON c.CustomerKey = s.CustomerKey
WHERE pc.EnglishProductCategoryName <> N'Components'
GROUP BY s.CustomerKey
HAVING COUNT(DISTINCT pc.EnglishProductCategoryName) =
  (SELECT COUNT(*) FROM dbo.DimProductCategory
   WHERE EnglishProductCategoryName <> N'Components')
ORDER BY s.CustomerKey;
```

Figure 6-8 shows the result.

	cid	cnm
1	11000	Yang, Jon
2	11001	Huang, Eugene
3	11003	Zhu, Christy
4	11025	Beck, Alejandro
5	11026	Sai, Harold
6	11027	Zhao, Jessie
7	11029	Moreno, Jimmy
8	11032	Stone, Denise

Figure 6-8. *Customers that purchased all product categories*

Adding the time is simple: expand the filter in the WHERE clause. To do the expansion, I joined the dbo.FactInternetSales table with the dbo.DimDate table, as shown in Listing 6-9.

Listing 6-9. Relational Division Limited by Time

```
SELECT s.CustomerKey AS cid,
  MIN(c.LastName + N', ' + c.FirstName) AS cnm
FROM dbo.FactInternetSales AS s
 INNER JOIN dbo.DimProduct AS p
  ON p.ProductKey = s.ProductKey
 INNER JOIN dbo.DimProductSubcategory AS ps
  ON ps.ProductSubcategoryKey = p.ProductSubcategoryKey
 INNER JOIN dbo.DimProductCategory AS pc
  ON pc.ProductCategoryKey = ps.ProductCategoryKey
 INNER JOIN dbo.DimCustomer AS c
  ON c.CustomerKey = s.CustomerKey
 INNER JOIN dbo.DimDate AS d
  ON d.DateKey = s.OrderDateKey
WHERE pc.EnglishProductCategoryName <> N'Components'
  AND YEAR(d.FullDateAlternateKey) = 2012
GROUP BY s.CustomerKey
HAVING COUNT(DISTINCT pc.EnglishProductCategoryName) =
  (SELECT COUNT(*) FROM dbo.DimProductCategory
    WHERE EnglishProductCategoryName <> N'Components')
ORDER BY s.CustomerKey;
```

Figure 6-9 shows the abbreviated results.

	cid	cnm
1	11117	Deng, April
2	11400	Raji, Franklin
3	11402	Cai, Kelli
4	11448	Patterson, Kyle
5	12124	Gill, Brandi
6	13059	Deng, Keith
7	16313	Russell, Hailey
8	16688	Gomez, Tabitha

Figure 6-9. *Customers who purchased from all categories in the year 2012*

Now we know which customers buy all product categories. These are very valuable customers. They are probably very loyal since they buy practically everything we sell. However, are they the best customers from the sales amount, sales quantity, or profit perspective?

Top Customers and Products

Let's start again with a formal definition. The *ABC analysis* is a categorization method in which an entity set of interest (e.g., customers, products) is divided into three categories, A, B, and C, in descending value of the target attribute, like sales amount or profit. A has the group with the highest value, the value of group B is lower than the value of group A, and group C has the lowest value. The following is the typical profit breakdown.

- A-items: 20% of all cases contribute to 70–80% of the profit

- B-items: 30% of all cases contribute to 15–25% of the profit

- C-items: 50% of all cases contribute to 5% of the profit

The *Pareto principle* is a special case of the ABC analysis. The Pareto principle says that 80% of the overall value comes from 20% of items. This is a very typical split that is done everywhere, not just in business. For example, when developing code, it is a common saying that you write 80% of the code 20% of the time, and you spend 80% of the time on the last 20% of the code.

For the ABC analysis, I used the code developed in Chapter 1. I calculated the frequencies and then used the cumulative percentage to find the cutting points, such as which products contribute 80% to the total sales. Listing 6-10 analyzes customers by the number of order lines, telling us who the most frequent customers are.

Listing 6-10. Analyzing Customers by Order Lines

```
WITH freqCTE AS
(
SELECT CustomerKey AS cid,
 COUNT(*) AS abf,
 CAST(ROUND(100. * (COUNT(*)) /
       (SELECT COUNT(*) FROM dbo.vInternetSales), 5)
       AS NUMERIC(8,5)) AS abp
FROM dbo.vInternetSales
```

```
GROUP BY CustomerKey
)
SELECT cid,
 abf,
 SUM(abf)
  OVER(ORDER BY abf DESC
        ROWS BETWEEN UNBOUNDED PRECEDING
        AND CURRENT ROW) AS cuf,
 abp,
 CAST(REPLICATE('*', ROUND(100 * abp, 0)) AS VARCHAR(50)) AS hst,
 SUM(abp)
  OVER(ORDER BY abf DESC
        ROWS BETWEEN UNBOUNDED PRECEDING
        AND CURRENT ROW) AS cup
FROM freqCTE
ORDER BY abf DESC;
```

Figure 6-10 shows the top ten customers based on the number of order lines. The top ten customers contribute nearly one percent of order lines. This is not valuable information.

	cid	abf	cuf	abp	hst	cup
1	11300	67	67	0.11470	**********	0.11470
2	11185	65	132	0.11127	**********	0.22597
3	11262	59	191	0.10100	*********	0.32697
4	11277	57	248	0.09758	*********	0.42455
5	11091	56	304	0.09587	*********	0.52042
6	11331	56	360	0.09587	*********	0.61629
7	11330	56	416	0.09587	*********	0.71216
8	11566	55	471	0.09416	********	0.80632
9	11223	54	525	0.09244	********	0.89876
10	11287	54	579	0.09244	********	0.99120

Figure 6-10. *Customers by the number of order lines*

Counting order lines does not reveal much about customers when using the demo data from the AdventureWorksDW2017 database. Let's try something else. I analyze the product models. Note that for a single model, you can have multiple products. In addition, instead of counting the order lines, I aggregate the sales amount, as shown in Listing 6-11.

Listing 6-11. Analyzing Product Models

```
WITH psalesCTE AS
(
SELECT ProductKey AS pid, SalesAmount AS sls
FROM dbo.vInternetSales
UNION
SELECT ProductKey, SalesAmount
FROM dbo.vResellerSales
),
psalsaggCTE AS
(
SELECT P.ModelName AS pmo,
 SUM(s.sls) AS sls
FROM psalesCTE AS s
 INNER JOIN dbo.DimProduct AS p
  ON s.pid = p.ProductKey
GROUP BY P.ModelName
)
SELECT pmo,
 ROUND(sls, 2) AS sls,
 RANK() OVER(ORDER BY sls DESC) AS rnk,
 SUM(ROUND(sls, 2)) OVER(ORDER BY sls DESC) runtot,
 ROUND( 100 *
  SUM(sls) OVER(ORDER BY sls DESC) /
  SUM(sls) OVER(), 2) AS runpct
FROM psalsaggCTE
ORDER BY sls DESC;
```

Figure 6-11 shows that the top ten product models contribute more than three quarters to the total sales amount. Now, this is a valuable piece of information.

	pmo	sls	rnk	runtot	runpct
1	Mountain-200	1354059.32	1	1354059.32	14.49
2	Mountain-100	1147202.90	2	2501262.22	26.77
3	Touring-1000	1141565.19	3	3642827.41	38.99
4	Road-250	956545.94	4	4599373.35	49.22
5	Road-650	817016.21	5	5416389.56	57.97
6	Road-350-W	493914.94	6	5910304.50	63.25
7	Touring-3000	388682.44	7	6298986.94	67.41
8	Road-550-W	355124.79	8	6654111.73	71.21
9	HL Mountain Frame	339539.28	9	6993651.01	74.85
10	HL Touring Frame	250142.35	10	7243793.36	77.52

Figure 6-11. *Top 10 products by sales amount*

We could now continue with analyzing top customers and products over specific years or other periods. However, this is an easy task, and I am sure you can do it yourself.

Let's switch to another group of analyses that deals with the loyalty of the customers.

Duration of Loyalty

In contemporary times, when we have a general perception that everything must change, we tend to forget about the importance of loyalty. For example, Internet and mobile network providers focus much more on new customers than on their existing ones. They give new customers all kinds of discounts and freebies, while loyal customers pay full price. As a result, these providers have raised a new breed of customers, who change providers when the discounts have ended. Because these customers are not paying full price, they are less valuable than the commonly forgotten loyal customers. But any local pub owner knows that the pub thrives due to the regular customers, not from the discount searchers.

When you use a subscription model, you know the exact date that a customer leaves you. When you do not use contracts, like in retail business, you depend on the occasional visits of your customers. In your database, you do not store the exact date that a customer abandoned you. You need to define the cutoff data yourself. For example, you can define that a customer who has not purchased anything in the last year is a lost customer.

Listing 6-12 lists the customers who did not purchase for more than a year and includes the number of days since their last purchase.

Listing 6-12. Finding Inactive Customers

```
SELECT s.CustomerKey AS cid,
 MIN(c.LastName + N', ' + c.FirstName) AS cnm,
 MAX(CAST(s.OrderDate AS DATE)) AS lod,
 DATEDIFF(day, MAX(s.OrderDate),
 (SELECT DATEADD(year, -1, MAX(OrderDate)) + 1
  FROM dbo.vInternetSales)) AS ddif
FROM dbo.vInternetSales AS s
 INNER JOIN dbo.DimCustomer AS c
  ON c.CustomerKey = s.CustomerKey
GROUP BY s.CustomerKey
HAVING MAX(s.OrderDate) <
 (SELECT DATEADD(year, -1, MAX(OrderDate)) + 1
  FROM dbo.vInternetSales)
ORDER BY lod;
```

Figure 6-12 shows the partial result.

	cid	cnm	lod	ddif
1	27601	Rogers, Sydney	2011-01-02	730
2	27612	Hill, Lucas	2011-01-03	729
3	27577	Cook, Patrick	2011-01-06	726
4	27666	Garcia, Alyssa	2011-01-06	726
5	25861	Cooper, Garrett	2011-01-06	726

Figure 6-12. *Customers inactive for more than a year*

You can also calculate the opposite from the number of days since the last purchase. This includes the number of days those customers were loyal. You could also use an arbitrary cutoff date. Listing 6-13 sets the cutoff date as February 28, 2013, and calculates the tenure days as the difference in days between the customer's first and last order date.

Listing 6-13. Calculating Tenure Days

```
SELECT i.CustomerKey AS cid,
 MIN(c.LastName + N', ' + c.FirstName) AS cnm,
 CAST(MIN(i.OrderDate) AS DATE) AS startDate,
 CAST(MAX(i.OrderDate) AS DATE) AS stopDate,
 DATEDIFF(day, MIN(i.OrderDate), MAX(i.OrderDate)) AS tenureDays
FROM dbo.vInternetSales AS i
 INNER JOIN dbo.DimCustomer AS c
  ON c.CustomerKey = i.CustomerKey
GROUP BY i.CustomerKey
HAVING MAX(i.OrderDate) < '20130301'
    AND DATEDIFF(day, MIN(i.OrderDate), MAX(i.OrderDate)) > 0
ORDER BY tenureDays DESC;
```

Figure 6-13 shows a few of the customers, sorted by descending tenure days.

	cid	cnm	startDate	stopDate	tenureDays
1	11002	Torres, Ruben	2011-01-07	2013-02-23	778
2	16482	Arthur, Tabitha	2011-01-11	2013-02-20	771
3	16514	Fernandez, Manuel	2011-01-10	2013-02-10	762
4	16740	Li, Elijah	2011-01-31	2013-02-14	745
5	16675	Arthur, Kristi	2011-02-03	2013-02-16	744

Figure 6-13. *Customers that left sorted by the tenure days*

Now we have developed the base queries for survival and hazard analyses.

Survival Analysis

In *survival analysis,* you try to estimate the *lifespan* of a particular population of interest, such as customers. It is also called the *time to event analysis.* You estimate the time that a customer starts to experience an event. A similar analysis is a *mean time between failure* (MTBF) analysis, where you estimate reliability.

When you are not dealing with a subscription model, a subject can enter and exit at any time in the study. Not every case experiences the event of interest, like churn or failure. In addition, besides a customer leaving you voluntarily, you might also have involuntary departures, such as when you decide to stop doing business with a customer—maybe because the customer is not paying regularly. You must have a variable that indicates the event of interest—a signal or a flag—that tells you whether the churn was voluntary or involuntary.

Listing 6-14 stores the customers who succumbed to risk after February 28, 2013, in a table and an artificial flag that makes the churn involuntary for one-third of those customers.

Listing 6-14. Storing the Former Customers to a Table

```
DROP TABLE IF EXISTS dbo.CustomerSurvival;
SELECT i.CustomerKey AS cId,
 MIN(c.LastName + N', ' + c.FirstName) AS cName,
 CAST(MIN(i.OrderDate) AS DATE) AS startDate,
 CAST(MAX(i.OrderDate) AS DATE) AS stopDate,
 DATEDIFF(day, MIN(i.OrderDate), MAX(i.OrderDate)) AS tenureDays,
 CASE WHEN i.CustomerKey % 3 = 0 THEN MIN(N'I') -- involuntary
      ELSE MIN(N'V')                            -- voluntary
 END AS stopReason
INTO dbo.CustomerSurvival
FROM dbo.vInternetSales AS i
 INNER JOIN dbo.DimCustomer AS c
  ON c.CustomerKey = i.CustomerKey
GROUP BY i.CustomerKey
HAVING MAX(i.OrderDate) < '20130301'
   AND DATEDIFF(day, MIN(i.OrderDate), MAX(i.OrderDate)) > 0;
```

Listing 6-15 adds the customers who are still active in this table. These are customers who made a purchase after February 28, 2013. I was not interested in new customers, so I filtered out all the customers who made their first purchase after the cutoff date. I also added a primary key constraint to the table.

Listing 6-15. Inserting Active Customers

```
INSERT INTO dbo.CustomerSurvival
SELECT i.CustomerKey AS cId,
 MIN(c.LastName + N', ' + c.FirstName) AS cName,
 CAST(MIN(i.OrderDate) AS DATE) AS startDate,
 NULL AS stopDate,
 DATEDIFF(day, MIN(i.OrderDate), '20130228') AS tenureDays,
 NULL AS stopReason
FROM dbo.vInternetSales AS i
 INNER JOIN dbo.DimCustomer AS c
  ON c.CustomerKey = i.CustomerKey
WHERE i.CustomerKey NOT IN
 (SELECT cId FROM dbo.CustomerSurvival)
GROUP BY i.CustomerKey
HAVING MIN(i.OrderDate) < '20130301';
ALTER TABLE dbo.customerSurvival ADD CONSTRAINT PK_CustSurv
 PRIMARY KEY (cId);
GO
```

Figure 6-14 shows a few random rows from the table. The customers who are still active have unknown values in the stopDate and stopReason columns.

	cId	cName	startDate	stopDate	tenureDays	stopReason
1	18701	Lal, Kenneth	2011-06-18	2013-02-08	601	V
2	22998	Washington, Lauren	2011-08-27	2012-12-28	489	I
3	18632	Tang, Russell	2013-02-20	NULL	8	NULL
4	23069	Gray, Anna	2013-02-14	NULL	14	NULL
5	24462	Fernandez, Donald	2012-11-17	NULL	103	NULL

Figure 6-14. *The content of the table for the survival analysis*

Many interesting analyses can be done with this analysis. Listing 6-16 filters the customers who first purchased before February 28, 2013. The code calculates the number of customers, the number of customers who succumbed to risk in the first year, the number of customers who succumbed to risk after one year, and the number and percentage of customers still active.

Listing 6-16. Survival Analysis

```
SELECT 365 AS tenureCutoff,
 COUNT(*) AS populationAtRisk,
 SUM(CASE WHEN tenureDays < 365 AND stopReason IS NOT NULL
     THEN 1 ELSE 0 END) AS succIn1Year,
 SUM(CASE WHEN tenureDays >= 365 AND stopReason IS NOT NULL
     THEN 1 ELSE 0 END) AS succAfter1Year,
 SUM(CASE WHEN tenureDays >= 365 AND stopReason IS NULL
     THEN 1 ELSE 0 END) AS numActive,
 CAST(
  ROUND(
   AVG(CASE WHEN tenureDays >= 365 AND stopReason IS NULL
       THEN 100.0 ELSE 0 END), 2)
  AS NUMERIC(5,2)) AS pctAcrive
FROM dbo.CustomerSurvival
WHERE startDate <= DATEADD(day, -365, '20130228');
GO
```

Figure 6-15 shows the result. Only nine customers succumbed to risk and became inactive in the first year; many more after one year. Yet, more than 91 percent of the customers who joined a year before the cutoff date are still active.

	tenureCutoff	populationAtRisk	succIn1Year	succAfter1Year	numActive	pctActive
1	365	2728	9	232	2487	91.17

Figure 6-15. *The result of the survival analysis*

We could continue with survival analysis by using different cutoff dates, different intervals of interest, such as calculating the customers who left after 18 months and so forth. But I want to switch to the view from the opposite side and analyze the customers who left.

Hazard Analysis

The *hazard function* (also called the *intensity function*) is the probability that the case experiences an event of interest within a specific time interval when the subject has survived until the beginning of that interval. The *risk* of the event at time t is the number of cases with the event in the interval beginning at time, *t*, divided by the product of the number of cases surviving at time *t* and the interval width. Again, this looks like a complex definition. But when you express the intention of the analysis, everything becomes quite simple. The query in Listing 6-17 calculates the hazard probability for the first year. Specifically, it calculates the number and the percentage of the customers who left within their first year.

Listing 6-17. Calculating the Hazard Probability for the First Year

```
SELECT 365 AS tenureCutoff,
 COUNT(*) AS populationAtRisk,
 SUM(CASE WHEN tenureDays <= 365 AND stopReason IS NOT NULL
     THEN 1 ELSE 0 END) AS succumbedToRisk,
 CAST(
  ROUND(
   AVG(CASE WHEN tenureDays <= 365 AND stopReason IS NOT NULL
       THEN 100.0 ELSE 0 END)
  , 2)
  AS NUMERIC(5,2)) AS pctSuccumbed
FROM dbo.CustomerSurvival;
```

Figure 6-16 shows the result. According to the data, the probability that a customer leaves within the first year of the customer's first purchase is less than three percent.

	tenureCutoff	populationAtRisk	succumbedToRisk	pctSuccumbed
1	365	6856	183	2.67

Figure 6-16. *Hazard probability for the first year*

The next thing is to calculate the hazard probability for every single possible value of the tenure days. This way, you can find the tenure days that are the riskiest. The query in Listing 6-18 does this calculation.

Listing 6-18. Finding Dangerous Tenure Lengths

```
SELECT tenureDays,
 COUNT(*) AS populationAtRisk,
 SUM(CASE WHEN stopReason IS NOT NULL
     THEN 1 ELSE 0 END) AS succumbedAtTenure
FROM dbo.CustomerSurvival
GROUP BY tenureDays
ORDER BY succumbedAtTenure DESC;
```

The output shown in Figure 6-17 lists, in descending order, the number of the customers who succumbed to risk at specific tenure days.

	tenureDays	populationAtRisk	succumbedAtTenure
1	628	9	5
2	171	13	4
3	612	13	4
4	609	11	4
5	200	12	4
6	498	8	3

Figure 6-17. *Critical values for the tenure days*

It does not look like there are any critical days in the tenure length. In a subscription model, you should get the critical lengths of the tenure related to the common lengths of the contracts.

Now let's write the master query for this section. The query calculates the population at risk: the number and cumulative percentage of customers who succumbed to risk for each length of the tenure (in days), and the number and percentage of the customers who survived each length of the tenure (in days). The query is shown in Listing 6-19.

Listing 6-19. Calculating Hazard and Survival Together

```
WITH tenCTE AS
(
SELECT tenureDays,
 COUNT(*) AS popAtRisk,
```

```
 SUM(CASE WHEN stopReason IS NOT NULL
     THEN 1 ELSE 0 END) AS succAtTenure
FROM dbo.CustomerSurvival
GROUP BY tenureDays
)
SELECT tenureDays,
 popAtRisk,
 SUM(popAtRisk) OVER() AS totPop,
 succAtTenure,
 SUM(succAtTenure)
  OVER(ORDER BY tenureDays
       ROWS BETWEEN UNBOUNDED PRECEDING
       AND CURRENT ROW) AS succUpToTenure,
 SUM(100.0 * succAtTenure)
  OVER(ORDER BY tenureDays
       ROWS BETWEEN UNBOUNDED PRECEDING
       AND CURRENT ROW) /
 SUM(popAtRisk) OVER() AS pctSuccUpToTenure,
 SUM(popAtRisk) OVER() -
 SUM(succAtTenure)
  OVER(ORDER BY tenureDays
       ROWS BETWEEN UNBOUNDED PRECEDING
       AND CURRENT ROW) AS survUpToTenure,
 100.0 -
 SUM(100.0 * succAtTenure)
  OVER(ORDER BY tenureDays
       ROWS BETWEEN UNBOUNDED PRECEDING
       AND CURRENT ROW) /
 SUM(popAtRisk) OVER() AS pctSurvUpToTenure
FROM tenCTE
ORDER BY tenureDays;
```

The partial result for the first 12 possible lengths of the tenure in days is shown in Figure 6-18.

tenureDays	popAtRisk	totPop	succAtTenure	succUpToTenure	pctSuccUpToTenure	survUpToTenure	pctSurvUpToTenure
0	27	6856	0	0	0.000000	6856	100.000000
1	41	6856	1	1	0.014585	6855	99.985415
2	39	6856	1	2	0.029171	6854	99.970829
3	43	6856	1	3	0.043757	6853	99.956243
4	40	6856	0	3	0.043757	6853	99.956243
5	49	6856	2	5	0.072928	6851	99.927072
6	42	6856	2	7	0.102100	6849	99.897900
7	49	6856	1	8	0.116686	6848	99.883314
8	44	6856	2	10	0.145857	6846	99.854143
9	29	6856	0	10	0.145857	6846	99.854143
10	28	6856	0	10	0.145857	6846	99.854143
11	42	6856	0	10	0.145857	6846	99.854143

Figure 6-18. *Hazard and survival for every single length of the tenure*

You can do further analyses of the customers who succumbed to risk and those that survived. For example, you could check the profit percentage for both groups of customers in different regions. I leave further analysis to you.

Conclusion

You saw some popular analyses. In this chapter, you learned about the exponential moving average, the ABC and Pareto analyses, the relational division problem, and the survival and hazard analyses.

I cleaned my AdventureWorksDW2017 database with the following code.

```
-- Clean up
USE AdventureWorksDW2017;
DROP TABLE IF EXISTS dbo.EMA;
DROP VIEW IF EXISTS dbo.vInternetSales;
DROP VIEW IF EXISTS dbo.vResellersales;
DROP TABLE IF EXISTS dbo.CustomerSurvival;
GO
```

You also saw how to use the exponential moving average for forecasting.

Perhaps you have not been satisfied with long-term forecasting when the forecasted value is a constant. In Chapter 7, I show how to do regression analysis with T-SQL. I also discuss other predictive algorithms and market basket analysis.

PART IV

Data Science

CHAPTER 7

Data Mining

Every web and/or retail shop wants to know which products customers tend to buy together. Trying to predict a target discrete or continuous variable with few input variables is important in practically every type of business. This chapter introduces some of the most popular algorithms implemented in T-SQL. You learn about the following.

- Linear regression

- Association rules

- Look-alike modeling

- Naïve Bayes algorithm

Demo Data

Let's start by introducing some demo data. For the market basket analysis, I used two views from the AdventureWorksDW2017 demo database: the `dbo.vAssocSeqLineItems` and the `dbo.vAssocSeqOrders` views. For better performance, I stored the data from the views in permanent tables. Listing 7-1 shows the first five rows from both views and then creates two permanent tables with primary keys with the data from both views.

Listing 7-1. Preparing the Data for Market Basket Analyses

```
USE AdventureWorksDW2017;
GO

-- Line items and orders
SELECT TOP 5 *
FROM dbo.vAssocSeqLineItems;
```

© Dejan Sarka 2021
D. Sarka, *Advanced Analytics with Transact-SQL*, https://doi.org/10.1007/978-1-4842-7173-5_7

```
SELECT TOP 5 *
FROM dbo.vAssocSeqOrders;
GO

-- Creating a permanent table from dbo.vAssocSeqLineItems
DROP TABLE IF EXISTS dbo.tAssocSeqLineItems;
SELECT *
INTO dbo.tAssocSeqLineItems
FROM dbo.vAssocSeqLineItems;
GO
-- PK
ALTER TABLE dbo.tAssocSeqLineItems ADD CONSTRAINT PK_LI
 PRIMARY KEY CLUSTERED (OrderNumber, LineNumber);
GO

-- Creating a permanent table from dbo.vAssocSeqOrders
DROP TABLE IF EXISTS dbo.tAssocSeqOrders;
SELECT *
INTO dbo.tAssocSeqOrders
FROM dbo.vAssocSeqOrders;
GO
-- PK
ALTER TABLE dbo.tAssocSeqOrders ADD CONSTRAINT PK_O
 PRIMARY KEY CLUSTERED (OrderNumber);
GO
```

Figure 7-1 shows the first rows from both views.

	OrderNumber	LineNumber	Model
1	SO61313	1	Road-350-W
2	SO61313	2	Cycling Cap
3	SO61313	3	Sport-100
4	SO61314	1	Hitch Rack - 4-Bike
5	SO61315	1	ML Road Tire

	OrderNumber	CustomerKey	Region	IncomeGroup
1	SO61313	11427	Europe	High
2	SO61314	11211	North America	Moderate
3	SO61315	15146	Pacific	High
4	SO61316	21187	Pacific	Moderate
5	SO61317	17355	Pacific	Low

Figure 7-1. *Data for market basket analyses*

The line items of the orders (the dbo.vAssocSeqLineItems view and the dbo.tAssocSeqLineItems table) have only three columns: order number, line number, and product model. This is all I needed for the basic market basket analysis. The orders view and the table created from the view (the dbo.tAssocSeqOrders view and the dbo.tAssocSeqOrders table) include columns that are useful for more in-depth analyses, such as regions or income.

For predictive modeling, I used the dbo.vTargetMail view. In this view, there is a target variable, BikeBuyer, which is a flag showing whether the customer purchased a bike. Other columns can be used to explain this decision and predict for a new customer is interested in bikes.

Listing 7-2 shows a couple of columns from the dbo.vTargetMail view.

Listing 7-2. Browsing vTargetMail View

```
SELECT CommuteDistance, Region,
 YearlyIncome, NumberCarsOwned,
 BikeBuyer
FROM dbo.vTargetMail;
```

Figure 7-2 shows the partial results.

	CommuteDistance	Region	YearlyIncome	NumberCarsOwned	BikeBuyer
1	0-1 Miles	North America	80000.00	0	1
2	10+ Miles	Europe	90000.00	3	0
3	2-5 Miles	North America	20000.00	2	0
4	10+ Miles	Pacific	90000.00	3	1
5	2-5 Miles	Europe	80000.00	2	0

Figure 7-2. *Few columns of the dbo.vTargetMail view*

You can see that some variables are discrete, and some are continuous. Also, some variables are strings, and some are numbers.

Linear Regression

Linear regression is sometimes called the "mother of all predictive algorithms." Whenever you are using continuous variables for a data science project, you use one of the regression algorithms at some point. Linear regression is the simplest regression algorithm. With linear regression, you show the dependent variable as a linear function of the independent variable, like the following formula shows.

$$Y = a + bX$$

There are more complex regression algorithms. *Multiple regression* allows a Y response variable to be modeled as a linear function of a multidimensional feature vector, which is a linear function of multiple input variables. In *polynomial regression,* you express the association with a formula where at least one independent variable is introduced in the equation as an nth order polynomial.

For the linear regression function, you need to calculate the *slope* and the *intercept* of the line. You can imagine that the two variables form a two-dimensional plane. The values of the two variables define the coordinates of the points in the plane. You are searching for a line that best fits all the points. To calculate the line, you use the deviations from the line, and the difference between the actual value for Y_i and the Y line value for every X_i. Some deviations are positive, and some are negative; the sum of all the deviations is zero for the best-fit line. To make the calculation possible, you need to square the deviations and then search for the line that minimizes the squared deviations. This way, you get the formulas for the slope and the intercept of the line.

$$Slope(Y) = \frac{\sum_{i=1}^{n}(X_i - \mu(X)) * (Y_i - \mu(Y))}{\sum_{i=1}^{n}(X_i - \mu(X))^2}$$

$$Intercept(Y) = \mu(Y) - Slope(Y) * \mu(X)$$

Figure 7-3 shows the linear relationship between the engine displacement in cubic centimeters and consumption in liters per 100 kilometers. The data is from the dbo.mtcars table, which you saw in Chapter 1. To get smaller numbers, I divided the displacement by 100. This way, the shape of the linear function is the same. But it is easier to manually calculate the consumption based on displacement if you want to check the formulas for the slope and the intercept calculated in Listing 7-3.

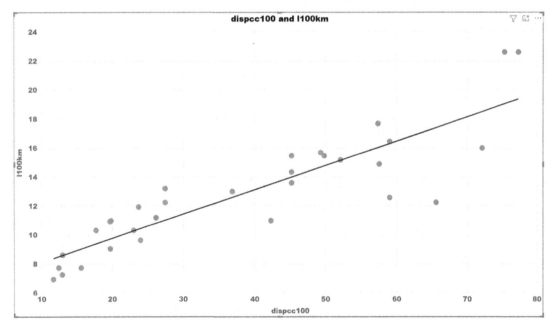

Figure 7-3. *Consumption as a function of displacement*

Listing 7-3 calculates the slope and the intercept of the linear line.

Listing 7-3. Calculating the Slope and the Intercept

```
WITH CoVarCTE AS
(
SELECT dispcc/100 as val1,
 AVG(dispcc/100) OVER () AS mean1,
 l100km AS val2,
 AVG(l100km) OVER() AS mean2
FROM dbo.mtcars
)
SELECT Slope =
        SUM((val1 - mean1) * (val2 - mean2))
        /SUM(SQUARE((val1 - mean1))),
      Intercept =
        MIN(mean2) - MIN(mean1) *
          (SUM((val1 - mean1)*(val2 - mean2))
           /SUM(SQUARE((val1 - mean1))))
FROM CoVarCTE;
```

The slope is approximately 0.17, and the intercept is 6.43. You can use these two numbers to estimate the consumption of any selected displacement.

Autoregression and Forecasting

When you deal with time series, you can use the previous values of the same variable for forecasting the future values with linear regression. This is called *autoregression.*

Chapter 6 forecasted with an exponential moving average (EMA). If you remember, the forecasted values quickly became constant. This chapter does the forecasting with EMA and linear regression (LR) using the same data to compare the two methods.

First, I stored the EMA values and the original values in a table, as shown in Listing 7-4. Note that I added the rows numbers in the table and a column named *LR* as a placeholder for the linear regression values. The code also adds a column named ValueType to indicate whether the values are actual or forecasted.

Listing 7-4. Storing EMA

```
DROP TABLE IF EXISTS dbo.EMA;
DECLARE @A AS FLOAT = 0.7;
WITH TSAggCTE AS
(
SELECT TimeIndex,
 CAST(SUM(1.0*Quantity*2) - 200 AS FLOAT) AS Quantity,
 DATEFROMPARTS(TimeIndex / 100, TimeIndex % 100, 1) AS DateIndex
FROM dbo.vTimeSeries
WHERE TimeIndex > 201012
GROUP BY TimeIndex
),
EMACTE AS
(
SELECT TimeIndex, Quantity,
  ROW_NUMBER() OVER (ORDER BY TimeIndex) - 1 AS Exponent
FROM TSAggCTE
)
SELECT
 ROW_NUMBER() OVER(ORDER BY TimeIndex) AS TimeRN,
 TimeIndex, Quantity,
 ROUND(
  SUM(CASE WHEN Exponent = 0 THEN 1
           ELSE @A
      END
      * POWER((1 - @A), -Exponent)
      * Quantity
  )
  OVER (ORDER BY TimeIndex)
  * POWER((1 - @A), Exponent)
 , 2) AS EMA,
 ROUND(CAST(0 AS FLOAT), 2) AS LR,
 CAST('Actual' AS CHAR(8)) AS ValueType
INTO dbo.EMA
FROM EMACTE;
GO
```

Listing 7-5 calculates the linear regression formula for the last 12 time points using the row number for the independent variable. It stores the slope and the intercept of the line in two variables. Then it uses these two values to update the LR column for the last 12 months. Then the code selects the values from the last row to calculate the EMA forecasts. Finally, the code calculates the forecasts with EMA and LR for the next six time points in a WHILE loop, inserts them in the table, and shows the results.

Listing 7-5. Forecasting with Autoregression

```
-- Linear regression - using last 12 points only
DECLARE @Slope AS FLOAT, @Intercept AS FLOAT;
WITH CoVarCTE AS
(
SELECT 1.0*TimeRN as val1,
 AVG(1.0*TimeRN) OVER () AS mean1,
 1.0*Quantity AS val2,
 AVG(1.0*Quantity) OVER() AS mean2
FROM dbo.EMA
WHERE TimeRN BETWEEN 25 AND 36
)
SELECT @Slope =
        SUM((val1 - mean1) * (val2 - mean2))
        /SUM(SQUARE((val1 - mean1))),
      @Intercept =
        MIN(mean2) - MIN(mean1) *
          (SUM((val1 - mean1)*(val2 - mean2))
            /SUM(SQUARE((val1 - mean1))))
FROM CoVarCTE;
-- Updating last 12 rows
UPDATE dbo.EMA
   SET LR = ROUND(TimeRN * @Slope + @Intercept, 2)
WHERE TimeRN BETWEEN 25 AND 36;
-- Selecting the last row
DECLARE @t AS INT,  @v AS FLOAT,
        @e AS FLOAT, @l AS FLOAT;
DECLARE @A AS FLOAT = 0.7, @r AS INT = 36;
```

```
SELECT @t = TimeIndex, @v = Quantity, @e = EMA
FROM dbo.EMA
WHERE TimeIndex =
      (SELECT MAX(TimeIndex) FROM dbo.EMA);
-- SELECT @t, @v, @e;
-- First forecast time point
SET @t = 201401;
SET @r = 37;
-- Forecasting in a loop
WHILE @t <= 201406
BEGIN
 SET @v = ROUND(@A * @v + (1 - @A) * @e, 2);
 SET @e = ROUND(@A * @v + (1 - @A) * @e, 2);
 SET @l = ROUND(@r * @Slope + @Intercept, 2);
 INSERT INTO dbo.EMA
 SELECT @r, @t, @v, @e, @l, 'Forecast';
 SET @t += 1;
 SET @r += 1;
END
SELECT TimeRN, TimeIndex,
 Quantity, EMA, LR,
 ValueType
FROM dbo.EMA
ORDER BY TimeIndex;
GO
```

Figure 7-4 shows the results of forecasting and the last few values.

	TimeRN	TimeIndex	Quantity	EMA	LR	ValueType
31	31	201307	788	812.31	802.5	Actual
32	32	201308	902	875.09	864.51	Actual
33	33	201309	832	844.93	926.52	Actual
34	34	201310	962	926.88	988.52	Actual
35	35	201311	1048	1011.66	1050.53	Actual
36	36	201312	1100	1073.5	1112.54	Actual
37	37	201401	1092.05	1086.49	1174.55	Forecast
38	38	201402	1090.38	1089.21	1236.55	Forecast
39	39	201403	1090.03	1089.78	1298.56	Forecast
40	40	201404	1089.95	1089.9	1360.57	Forecast
41	41	201405	1089.94	1089.93	1422.57	Forecast
42	42	201406	1089.94	1089.94	1484.58	Forecast

Figure 7-4. *Forecasted values with EMA and with LR*

Figure 7-5 shows the actual values and the forecasted values graphically. You can see the values, the exponential moving averages, and the values calculated with linear regression.

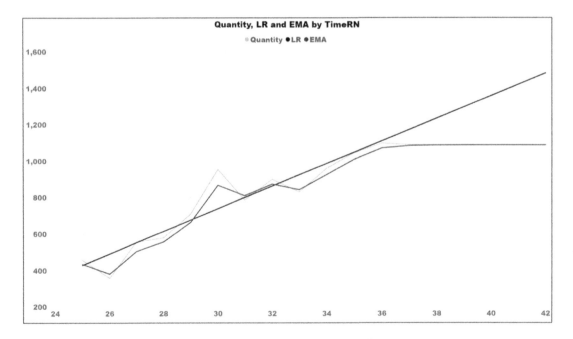

Figure 7-5. *Comparing forecasting with ema and with LR*

Which forecasted values—the EMA values or the LR values—are better? Usually, EMA is better for short-term forecasting, while linear regression is better for long-term forecasting. You might decide to use both methods.

Association Rules

Market basket analysis tries to find the relationships between products. Can a sale of one product raise or lower the probability of selling another product? You analyze the products sold together in a single transaction and in a single market basket. You try to express the relationships between products with *association rules.*

In the association rules algorithm, you consider each attribute in a value pair (such as product—ice cream) as an *item*. An *itemset* is a combination of items in a single transaction. You must scan through the dataset to find the itemsets that tend to appear in many transactions. Itemsets can have two or more items. You can check the itemsets with a single item. *Support* is the number of rows with the itemset as an absolute number or divided by the total number of rows and expressed as a percentage.

Listing 7-6 checks the most popular itemsets with a single item by counting the product models.

Listing 7-6. Checking itemsets with a Single Item

```
SELECT Model, COUNT(*) AS Support
FROM dbo.tAssocSeqLineItems
GROUP BY Model
ORDER BY Support DESC;
```

Figure 7-6 shows few most popular products.

	Model	Support
1	Sport-100	3782
2	Water Bottle	2489
3	Patch kit	1830
4	Mountain Tire Tube	1776
5	Mountain-200	1470
6	Road Tire Tube	1374

Figure 7-6. *Most popular products*

You can see, for example, that the water bottle is the second most popular product.

Starting from the Negative Side

Negative associations might be just as important as positive ones. I continue the market basket analysis by finding the products that have been purchased only once in a single order by a customer, meaning that the customers bought that product and never bought anything else. Listing 7-7 finds those products that tend to have a negative impact on associations.

Listing 7-7. Finding "Negative" Products

```
WITH mdlCTE AS
(
SELECT o.CustomerKey AS cid,
 MIN(i.Model) AS mdl,
 COUNT(DISTINCT i.OrderNumber) AS distor,
 COUNT(DISTINCT i.Model) AS distit
FROM dbo.tAssocSeqLineItems AS i
 INNER JOIN dbo.tAssocSeqOrders AS o
  ON i.OrderNumber = o.OrderNumber
GROUP BY o.CustomerKey
HAVING
 COUNT(DISTINCT i.OrderNumber) = 1 AND
 COUNT(DISTINCT i.Model) = 1
)
SELECT mdl AS Product, COUNT(*) AS Cnt
FROM mdlCTE
GROUP BY mdl
ORDER BY cnt DESC;
```

Figure 7-7 shows the partial result.

	Product	Cnt
1	Water Bottle	179
2	Women's Mountain Shorts	140
3	Road-750	120
4	Mountain-200	119
5	Road-350-W	95
6	Mountain Tire Tube	82

Figure 7-7. The "negative" products

Although the water bottle is the second-most popular product, it is a product that tends to be sold alone. Let's use the attributes from the order table for further analysis of these "negative" products. Listing 7-8 pivots the product over the region and income groups.

Listing 7-8. Further Analysis of the "Negative" Products

```
-- Pivoting by region
WITH mdlCTE AS
(
SELECT MIN(o.Region) AS reg,
 MIN(i.Model) AS mdl,
 COUNT(DISTINCT i.Model) AS distit
FROM dbo.tAssocSeqLineItems AS i
 INNER JOIN dbo.tAssocSeqOrders AS o
  ON i.OrderNumber = o.OrderNumber
GROUP BY o.CustomerKey
HAVING
 COUNT(DISTINCT i.OrderNumber) = 1 AND
 COUNT(DISTINCT i.Model) = 1
)
SELECT mdl AS Model,
 ISNULL([Europe], 0) AS Europe,
 ISNULL([North America], 0) AS NorthAm,
 ISNULL([Pacific], 0) AS Pacific
FROM mdlCTE
 PIVOT (SUM(distit) FOR reg
    IN ([Europe], [North America], [Pacific])) AS p
ORDER BY ISNULL([Europe], 0) +
         ISNULL([North America], 0) +
         ISNULL([Pacific], 0) DESC;
-- Pivoting by income group
WITH mdlCTE AS
(
SELECT MIN(o.IncomeGroup) AS reg,
 MIN(i.Model) AS mdl,
 COUNT(DISTINCT i.Model) AS distit
```

```
FROM dbo.tAssocSeqLineItems AS i
 INNER JOIN dbo.tAssocSeqOrders AS o
  ON i.OrderNumber = o.OrderNumber
GROUP BY o.CustomerKey
HAVING
 COUNT(DISTINCT i.OrderNumber) = 1 AND
 COUNT(DISTINCT i.Model) = 1
)
SELECT mdl AS Model,
 ISNULL([Low], 0) AS LowIncome,
 ISNULL([Moderate], 0) AS ModerateIncome,
 ISNULL([High], 0) AS HighIncome
FROM mdlCTE
 PIVOT (SUM(distit) FOR reg
    IN ([Low], [Moderate], [High])) AS p
ORDER BY ISNULL([Low], 0) +
         ISNULL([Moderate], 0) +
         ISNULL([High], 0) DESC;
```

Figure 7-8 shows the partial result.

	Model	Europe	NorthAm	Pacific
1	Water Bottle	45	114	20
2	Women's Mountain Shorts	11	106	23
3	Road-750	33	44	43
4	Mountain-200	34	55	30
5	Road-350-W	26	30	39

	Model	LowIncome	ModerateIncome	HighIncome
1	Water Bottle	47	55	77
2	Women's Mountain Shorts	19	63	58
3	Road-750	57	33	30
4	Mountain-200	15	21	83
5	Road-350-W	26	48	21

Figure 7-8. *The "negative" products over region and income group*

Before making conclusions about how the water bottle has a very negative impact on selling other products, remember that this is a generally very popular product. Therefore, it might frequently be in an itemset with other products.

Frequency of Itemsets

The next step is to calculate the frequency of itemsets with two products. Listing 7-9 shows two different solutions for this task. In the first solution, I used a join of the line items table based on the equality of the order number from both sides and the inequality of the product models in the same order. I used the greater-than operator for the product models because I wanted to get a pair of products only once. The second solution uses the CROSS APPLY operator for the same task.

Listing 7-9. Finding Itemsets with Two Items

```
-- Using JOIN
SELECT t1.Model AS Model1,
 t2.Model AS Model2,
 COUNT(*) AS Support
FROM dbo.tAssocSeqLineItems AS t1
 INNER JOIN dbo.tAssocSeqLineItems AS t2
  ON t1.OrderNumber = t2.OrderNumber
     AND t1.Model > t2.Model
GROUP BY t1.Model, t2.Model
ORDER BY Support DESC;
-- Using APPLY
WITH Pairs_CTE AS
(
SELECT t1.OrderNumber,
 t1.Model AS Model1,
 t2.Model2
FROM dbo.tAssocSeqLineItems AS t1
 CROSS APPLY
  (SELECT Model AS Model2
   FROM dbo.tAssocSeqLineItems
```

```
    WHERE OrderNumber = t1.OrderNumber
        AND Model > t1.Model) AS t2
)
SELECT Model1, Model2, COUNT(*) AS Support
FROM Pairs_CTE
GROUP BY Model1, Model2
ORDER BY Support DESC;
```

Figure 7-9 shows the partial result of the first query only. Of course, the result of the second query is completely the same.

	Model1	Model2	Support
1	Water Bottle	Mountain Bottle Cage	993
2	Water Bottle	Road Bottle Cage	892
3	Sport-100	Mountain Tire Tube	747
4	Water Bottle	Sport-100	649
5	Mountain Tire Tube	HL Mountain Tire	551

Figure 7-9. *The most popular itemsets with two items*

Now you can see that water bottles can have a positive impact. They are frequently bought with bottle cages.

For further analysis, I used the query that uses the APPLY operator. It is simple and intuitive to expand this query to find the itemsets with three items. You only need to use the result of the itemset with two items and add the third item with another APPLY operator, like Listing 7-10 shows.

Listing 7-10. Finding itemsets with Three Items

```
WITH Pairs_CTE AS
(
SELECT t1.OrderNumber,
 t1.Model AS Model1,
 t2.Model2
FROM dbo.tAssocSeqLineItems AS t1
 CROSS APPLY
  (SELECT Model AS Model2
    FROM dbo.tAssocSeqLineItems
```

```
    WHERE OrderNumber = t1.OrderNumber
      AND Model > t1.Model) AS t2
),
Triples_CTE AS
(
SELECT t2.OrderNumber,
 t2.Model1,
 t2.Model2,
 t3.Model3
FROM Pairs_CTE AS t2
 CROSS APPLY
  (SELECT Model AS Model3
    FROM dbo.tAssocSeqLineItems
    WHERE OrderNumber = t2.OrderNumber
      AND Model > t2.Model1
      AND Model > t2.Model2) AS t3
)
SELECT Model1, Model2, Model3, COUNT(*) AS Support
FROM Triples_CTE
GROUP BY Model1, Model2, Model3
ORDER BY Support DESC;
GO
```

Figure 7-10 shows that the water bottle goes well with other products.

	Model1	Model2	Model3	Support
1	Mountain Bottle Cage	Mountain-200	Water Bottle	343
2	Mountain Bottle Cage	Sport-100	Water Bottle	281
3	Road Bottle Cage	Road-750	Water Bottle	278
4	HL Mountain Tire	Mountain Tire Tube	Sport-100	233
5	Road Bottle Cage	Sport-100	Water Bottle	223

Figure 7-10. *The most popular itemsets with three items*

Support is not the only measure for the value of the association rules. Oh! Let's stop for a second. Which rules? So far, we have only searched for combinations of the items.

Association Rules

The association rules are expressed as sentences with a condition and a consequence, like "if product A, then product B." Listing 7-11 expresses the combinations of two items as rules. Please note that this time I returned every itemset twice, expressed as two rules wherein I have both products used in the condition and once in the consequence of the rule.

Listing 7-11. Rules for the itemsets of two items

```
WITH Pairs_CTE AS    -- All possible pairs
(
SELECT t1.OrderNumber,
 t1.Model AS Model1,
 t2.Model2
FROM dbo.tAssocSeqLineItems AS t1
 CROSS APPLY
  (SELECT Model AS Model2
    FROM dbo.tAssocSeqLineItems
    WHERE OrderNumber = t1.OrderNumber
      AND Model <> t1.Model) AS t2
)
SELECT Model1 + N' ---> ' + Model2 AS theRule,
 Model1, Model2, COUNT(*) AS Support
FROM Pairs_CTE
GROUP BY Model1, Model2
ORDER BY Support DESC;
```

Figure 7-11 shows the result. Each pair of items in an itemset appears twice in the results with the same support.

	theRule	Model1	Model2	Support
1	Water Bottle ---> Mountain Bottle Cage	Water Bottle	Mountain Bottle Cage	993
2	Mountain Bottle Cage ---> Water Bottle	Mountain Bottle Cage	Water Bottle	993
3	Road Bottle Cage ---> Water Bottle	Road Bottle Cage	Water Bottle	892
4	Water Bottle ---> Road Bottle Cage	Water Bottle	Road Bottle Cage	892
5	Mountain Tire Tube ---> Sport-100	Mountain Tire Tube	Sport-100	747
6	Sport-100 ---> Mountain Tire Tube	Sport-100	Mountain Tire Tube	747

Figure 7-11. *The rules for the itemsets with two items*

Now every pair of items expressed has two rules. The question is whether there is a difference between the two rules in a pair. Is one rule somehow better than the other one? Let's use the water bottle example again. It appears in many combinations, especially with a bottle cage; however, it also appears alone many times—more times than the bottle cage appears alone. You might be more confident that a customer would buy a water bottle if she buys a bottle cage than the customer buying a bottle cage when she buys a water bottle.

The *confidence* of an association rule is defined as the support for the combination divided by the support for the condition.

Listing 7-12 also calculates confidence and then orders the rules by support and then by confidence, both times in descending order. I calculated the frequency of the itemsets with a single item in an additional common table expression. I also calculated support as a percentage. For this calculation, I needed the number of orders, which I added to the rules through a cross join with a subquery that returns this scalar value.

Listing 7-12. Finding Association Rules with Support and Confidence

```
WITH Pairs_CTE AS    -- All possible pairs
(
SELECT t1.OrderNumber,
 t1.Model AS Model1,
 t2.Model2
FROM dbo.tAssocSeqLineItems AS t1
 CROSS APPLY
  (SELECT Model AS Model2
    FROM dbo.tAssocSeqLineItems
    WHERE OrderNumber = t1.OrderNumber
      AND Model <> t1.Model) AS t2
),
rulesCTE AS
(
SELECT Model1 + N' ---> ' + Model2 AS theRule,
 Model1, Model2, COUNT(*) AS Support
FROM Pairs_CTE
GROUP BY Model1, Model2
),
```

```
cntModelCTE AS
(
SELECT Model,
 COUNT(DISTINCT OrderNumber) AS ModelCnt
FROM dbo.tAssocSeqLineItems
GROUP BY Model
)
SELECT r.theRule,
 r.Support,
 CAST(100.0 * r.Support / a.numOrders AS NUMERIC(5, 2))
  AS SupportPct,
 CAST(100.0 * r.Support / c1.ModelCnt AS NUMERIC(5, 2))
  AS Confidence
FROM rulesCTE AS r
 INNER JOIN cntModelCTE AS c1
  ON r.Model1 = c1.Model
 CROSS JOIN (SELECT COUNT(DISTINCT OrderNumber)
             FROM dbo.tAssocSeqLineItems) AS a(numOrders)
ORDER BY Support DESC, Confidence DESC;
```

Let's check the result in Figure 7-12.

	theRule	Support	SupportPct	Confidence
1	Mountain Bottle Cage ---> Water Bottle	993	7.63	83.03
2	Water Bottle ---> Mountain Bottle Cage	993	7.63	39.90
3	Road Bottle Cage ---> Water Bottle	892	6.86	89.29
4	Water Bottle ---> Road Bottle Cage	892	6.86	35.84
5	Mountain Tire Tube ---> Sport-100	747	5.74	42.06
6	Sport-100 ---> Mountain Tire Tube	747	5.74	19.75
7	Water Bottle ---> Sport-100	649	4.99	26.07
8	Sport-100 ---> Water Bottle	649	4.99	17.16

Figure 7-12. *Association rules with confidence*

Now you can see that some rules are better than others. Although the water bottle is a very popular product in combinations, it also has a negative effect on the probability that there will be another product in the basket when a water bottle is in it.

There is another important measure for the quality of the association rules: *lift*, sometimes called *importance*. Lift tells you how often it is likely that a customer buys product B when product A is in the basket compared to the likelihood of buying product B alone. Lift is defined in the following formula.

$$Lift(A,B) = \frac{P(A \wedge B)}{P(A) * P(B)}$$

The following is the interpretation of lift.

- If lift = 1, A and B are independent items.

- If lift > 1, A and B are positively correlated.

- If lift < 1, A and B are negatively correlated.

Listing 7-13 adds lift to the rules.

Listing 7-13. Adding Lift

```
WITH Pairs_CTE AS    -- All possible pairs
(
SELECT t1.OrderNumber,
 t1.Model AS Model1,
 t2.Model2
FROM dbo.tAssocSeqLineItems AS t1
 CROSS APPLY
  (SELECT Model AS Model2
   FROM dbo.tAssocSeqLineItems
   WHERE OrderNumber = t1.OrderNumber
     AND Model <> t1.Model) AS t2
),
rulesCTE AS
(
SELECT Model1 + N' ---> ' + Model2 AS theRule,
 Model1, Model2, COUNT(*) AS Support
FROM Pairs_CTE
GROUP BY Model1, Model2
),
```

```
cntModelCTE AS
(
SELECT Model,
 COUNT(DISTINCT OrderNumber) AS ModelCnt
FROM dbo.tAssocSeqLineItems
GROUP BY Model
)
SELECT r.theRule,
 r.Support,
 CAST(100.0 * r.Support / a.numOrders AS NUMERIC(5, 2))
  AS SupportPct,
 CAST(100.0 * r.Support / c1.ModelCnt AS NUMERIC(5, 2))
  AS Confidence,
 CAST((1.0 * r.Support / a.numOrders) /
  ((1.0 * c1.ModelCnt / a.numOrders) *
   (1.0 * c2.ModelCnt / a.numOrders)) AS NUMERIC(5, 2))
  AS Lift
FROM rulesCTE AS r
 INNER JOIN cntModelCTE AS c1
  ON r.Model1 = c1.Model
 INNER JOIN cntModelCTE AS c2
  ON r.Model2 = c2.Model
 CROSS JOIN (SELECT COUNT(DISTINCT OrderNumber)
            FROM dbo.tAssocSeqLineItems) AS a(numOrders)
ORDER BY Support DESC, Confidence DESC;
```

Figure 7-13 shows the result, ordered the same way as in Figure 7-12, by support and confidence in descending order.

	theRule	Support	SupportPct	Confidence	Lift
1	Mountain Bottle Cage ---> Water Bottle	993	7.63	83.03	4.34
2	Water Bottle ---> Mountain Bottle Cage	993	7.63	39.90	4.34
3	Road Bottle Cage ---> Water Bottle	892	6.86	89.29	4.67
4	Water Bottle ---> Road Bottle Cage	892	6.86	35.84	4.67
5	Mountain Tire Tube ---> Sport-100	747	5.74	42.06	1.45
6	Sport-100 ---> Mountain Tire Tube	747	5.74	19.75	1.45
7	Water Bottle ---> Sport-100	649	4.99	26.07	0.90
8	Sport-100 ---> Water Bottle	649	4.99	17.16	0.90

Figure 7-13. *Association Rules with Support, confidence, and Lift*

You can immediately see that the rules that initially seemed very important do not seem desirable. For example, the Water Bottle ➤ Sport-100 rule has a lift lower than 1, meaning that the association between these two products is negative. You can execute the query from Listing 7-13 again with the changed ORDER BY clause. If you order the result by lift, you see that the lift for some rules is very high, while others are extremely low. However, when lift is low, confidence is usually low.

If you analyze the market basket for a retail store, you are probably finished with the analysis when you find the rules with confidence. You cannot find the order of the products in a basket because order becomes mixed at checkout with the cashier. However, with Internet sales, you can check the *order* of the products in a basket. The demo data I used is about Internet sales. The sequence of the products in an order is defined by item line numbers. It is simple to add the order in an analysis. Here, I just needed to change the condition that the order line number from the left is lower than the order line number from the right in the same order when I find pairs of items, as shown in Listing 7-14. I did not calculate lift because the list is independent of the order of the items in the itemsets.

Listing 7-14. Analyzing Association Rules with Order of Items

```
WITH Pairs_CTE AS    -- All possible pairs
(
SELECT t1.OrderNumber,
 t1.Model AS Model1,
 t2.Model2
FROM dbo.tAssocSeqLineItems AS t1
 CROSS APPLY
```

```
  (SELECT Model AS Model2
   FROM dbo.tAssocSeqLineItems
   WHERE OrderNumber = t1.OrderNumber
     AND t1.LineNumber < LineNumber      -- sequence
     AND Model <> t1.Model) AS t2
),
rulesCTE AS
(
SELECT Model1 + N' ---> ' + Model2 AS theRule,
 Model1, Model2, COUNT(*) AS Support
FROM Pairs_CTE
GROUP BY Model1, Model2
),
cntModelCTE AS
(
SELECT Model,
 COUNT(DISTINCT OrderNumber) AS ModelCnt
FROM dbo.tAssocSeqLineItems
GROUP BY Model
)
SELECT r.theRule,
 r.Support,
 CAST(100.0 * r.Support / a.numOrders AS NUMERIC(5, 2))
  AS SupportPct,
 CAST(100.0 * r.Support / c1.ModelCnt AS NUMERIC(5, 2))
  AS Confidence
FROM rulesCTE AS r
 INNER JOIN cntModelCTE AS c1
  ON r.Model1 = c1.Model
 CROSS JOIN (SELECT COUNT(DISTINCT OrderNumber)
             FROM dbo.tAssocSeqLineItems) AS a(numOrders)
ORDER BY Support DESC, Confidence DESC;
```

Figure 7-14 shows the result.

	theRule	Support	SupportPct	Confidence
1	Mountain Tire Tube ---> Sport-100	747	5.74	42.06
2	Mountain Bottle Cage ---> Water Bottle	737	5.67	61.62
3	Water Bottle ---> Sport-100	649	4.99	26.07
4	Road Bottle Cage ---> Water Bottle	540	4.15	54.05
5	Road Tire Tube ---> Sport-100	517	3.98	37.63
6	Mountain Tire Tube ---> Patch kit	445	3.42	25.06

Figure 7-14. *Association rules with order of items*

When paired with another product, the water bottle usually goes in the basket later than the other product.

You can ask yourself another interesting question. Are there products that typically begin the shopping sequence and another set of products that usually end the shopping sequence? You can easily identify these products by adding two virtual products at the beginning and the end of the sequences. In our example, the order line number defines the sequence of the products in the market basket. The line number always starts with 1; the highest line number differs from order to order. However, there is no need to have sequences without gaps; the line number for the virtual product inserted at the end of each sequence, as the last line item on each order, must only have the line number higher than any other products in the same order. I calculated the minimal and the maximal line number with the following query.

```
SELECT MIN(LineNumber) as minLn,
 MAX(LineNumber) AS maxLn
FROM dbo.tAssocSeqLineItems;
```

The query shows that the minimal line number is 1, and the maximal line number is 8. Next, I inserted two virtual products on each order. The product named *Begin* has line number 0, and the product named *End* has line number 9, which is higher than the highest line number on any order. I inserted these two products with the following code.

```
INSERT INTO dbo.tAssocSeqLineItems
 (OrderNumber, LineNumber, Model)
SELECT DISTINCT OrderNumber, 0, N'Begin'
FROM dbo.tAssocSeqLineItems
UNION
SELECT DISTINCT OrderNumber, 9, N'End'
FROM dbo.tAssocSeqLineItems;
```

239

Listing 7-14 is executed again to find the association rules with the sequence. The partial results are shown in Figure 7-15.

	theRule	Support	SupportPct	Confidence
1	Begin ---> End	13006	100.00	100.00
2	Sport-100 ---> End	3782	29.08	100.00
3	Begin ---> Sport-100	3782	29.08	29.08
4	Water Bottle ---> End	2489	19.14	100.00
5	Begin ---> Water Bottle	2489	19.14	19.14
6	Patch kit ---> End	1830	14.07	100.00
7	Begin ---> Patch kit	1830	14.07	14.07
8	Mountain Tire Tube ---> End	1776	13.66	100.00
9	Begin ---> Mountain Tire Tube	1776	13.66	13.66
55	Begin ---> Bike Wash	524	4.03	4.03
56	Road Tire Tube ---> Sport-1...	517	3.98	37.63
57	LL Mountain Tire ---> End	499	3.84	100.00
58	Begin ---> LL Mountain Tire	499	3.84	3.84
59	HL Road Tire ---> End	461	3.54	100.00
60	Begin ---> HL Road Tire	461	3.54	3.54

Figure 7-15. *Sequences with Begin and End virtual products*

The Begin ➤ End rule is at the top of the list since these two virtual products now appear in every rule. Also, the rules with the Begin and End virtual products in combination with other products are generally very high on this list, ordered by support and confidence. At the beginning of the list are products that are usually sold alone, since they always appear in two rows, in two pairs of items, once with the Begin and once with the End virtual product. But later, the Bike Wash product frequently starts the sequence but does not end it. Nevertheless, with all the measures introduced in this section, you should have a pretty good understanding of the associations between products.

Look-Alike Modeling

Let's discuss *look-alike modeling* with the help of the *decision tree* algorithm (DT), which is a classification and prediction method. You must have a discrete target variable, probably a flag. Recall the `dbo.vTargetMail` view I introduced at the beginning of this chapter. There is such a target variable in the view—the `BikeBuyer` variable. This is a flag denoting whether a customer has purchased a bike in the past. Then you need a few discrete input variables. The algorithm uses these input features to explain the states of the target variable. Then you can use the knowledge you gained from learning about predictions of the target variable on a new dataset.

The DT algorithm does *recursive partitioning*. Initially, you have the whole dataset. The distribution of the target variable is known. The DT algorithm uses every input variable for splitting the data into groups of the input variable and measures the distribution of the target variable in these groups. It keeps the split over the variable that produced the cleanest distribution of the target variable. For example, let's say that the initial distribution of the target variable was 50/50. After a split over two different variables (for simplicity, with only two possible states), you get the following distributions in the groups: 60/40 and 38/62 in the first input variable, and 30/70 and 75/25 in the second input variable. The second input variable made the distribution of the target variable cleaner. Then the DT algorithm continues the splitting process using other input variables to split the groups into subgroups until a stopping condition is met.

A simple stopping condition is the number of cases, or rows, in subgroups after the split. If the number is too small, the split is useless. The whole process of building the model can be represented graphically in a tree. An example of a DT on the `dbo.vTargetMail` data is shown in Figure 7-16.

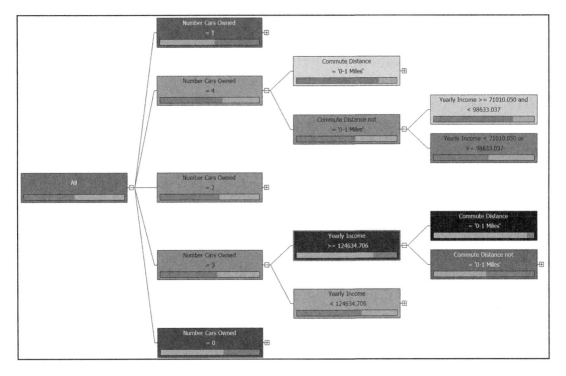

Figure 7-16. *Decision tree*

Figure 7-16 shows that the first split was done by the number of cars owned. There are subgroups for yearly income, commute distance, and so on. The process of building a tree is called *training* the model. Once the tree is built, it is easy to understand what drives customers to buy a bike. In addition, when you get a new customer with known input variables, you can classify the customer in the appropriate subgroup and then use the distribution of the target variable in that subgroup for predictions.

Training and Test Data Sets

Implementing an algorithm with recursion is not efficient in T-SQL. However, if we knew which variables had the most impact on the target variable, we could find the appropriate subgroup for a new case immediately with simple JOIN or APPLY operators. This would be very efficient. You could do classifications and predictions on very large datasets.

You can always use your experience and knowledge to select appropriate input variables. But you can also check which input variables have the biggest influence on the target variable by checking the associations between pairs of variables, one input and target at a time. I explained how to check the associations between pairs of variables in Chapter 2.

Figure 7-17 graphically shows associations between three input variables and the target variable. Specifically, it shows the distribution of the BikeBuyer target variable in the classes of the following input variables: Region, NumberCarsOwned, and CommuteDistance.

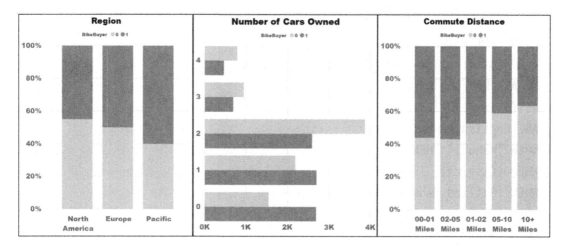

Figure 7-17. *Associations between the target and input variables*

The distribution of the target variable varies in the input variables' classes.

Next, let's develop look-alike modeling (LAM).

To get better analyses, you should do some basic data preparation. You can discretize continuous variables, change the classes of discrete variables, make character variables from numerics, and more. The query in Listing 7-14 is an example.

Listing 7-15. Preparing the Data

```
SELECT
 CAST(
 CASE EnglishEducation
  WHEN 'Partial High School' THEN '1'
  WHEN 'High School' THEN '2'
```

```
 WHEN 'Partial College' THEN '3'
 WHEN 'Bachelors' THEN '4'
 WHEN 'Graduate Degree' THEN '5'
 ELSE '0'          -- Handling possible NULLs
END AS CHAR(1)) AS Education,
CAST(
CASE CommuteDistance
 WHEN '0-1 Miles' THEN '1'
 WHEN '1-2 Miles' THEN '2'
 WHEN '2-5 Miles' THEN '3'
 WHEN '5-10 Miles' THEN '4'
 WHEN '10+ Miles' THEN '5'
 ELSE '0'          -- Handling possible NULLs
END AS CHAR(1)) AS CommDist,
CAST(
CASE EnglishOccupation
 WHEN 'Manual' THEN '1'
 WHEN 'Clerical' THEN '2'
 WHEN 'Skilled Manual' THEN '3'
 WHEN 'Professional' THEN '4'
 WHEN 'Management' THEN '5'
 ELSE '0'          -- Handling possible NULLs
END AS CHAR(1)) AS Occupation,
CAST(
 CASE Region
  WHEN 'Europe' THEN '1'
  WHEN 'North America' THEN '2'
  WHEN 'Pacific' THEN '3'
  ELSE '0'          -- Handling possible NULLs
 END AS CHAR(1)) AS Reg,
CAST(NTILE(5) OVER(ORDER BY Age) AS CHAR(1)) AS AgeEHB,
CAST(NTILE(5) OVER(ORDER BY YearlyIncome) AS CHAR(1)) AS IncEHB,
CAST(ISNULL(TotalChildren, 0) AS CHAR(1)) AS Children,
CAST(
 CASE NumberCarsOwned
```

```
  WHEN 0 THEN '1'
  WHEN 1 THEN '1'
  WHEN 2 THEN '2'
  ELSE '3'
 END AS CHAR(1)) AS Cars,
 *
FROM dbo.vTargetMail;
```

The query uses the NTILE() function to discretize the Age and YearlyIncome variables and the CASE expression to change the classes of discrete variables. After the changes, the values were a single character for each input variable.

Now let's concatenate all the input variables in a single string. This string is called the *grouping factor*. The process is shown in Listing 7-16. It uses the grouping factor for the GROUP BY clause and counts the rows in each group.

Listing 7-16. Calculating the Grouping Factor

```
WITH gfCTE AS
(
SELECT
 CAST(
 CASE NumberCarsOwned
  WHEN 0 THEN '1'
  WHEN 1 THEN '1'
  WHEN 2 THEN '2'
  ELSE '3'
 END AS CHAR(1)) +
 CAST(
 CASE CommuteDistance
  WHEN '0-1 Miles' THEN '1'
  WHEN '1-2 Miles' THEN '2'
  WHEN '2-5 Miles' THEN '2'
  WHEN '5-10 Miles' THEN '3'
  WHEN '10+ Miles' THEN '3'
  ELSE '0'
 END AS CHAR(1)) +
```

```
CAST(
 CASE Region
  WHEN 'Europe' THEN '1'
  WHEN 'North America' THEN '2'
  WHEN 'Pacific' THEN '3'
  ELSE '0'
 END AS CHAR(1)) +
CAST(NTILE(3) OVER(ORDER BY YearlyIncome) AS CHAR(1))
 AS GF,     -- grouping factor
 *
FROM dbo.vTargetMail
)
SELECT GF, COUNT(*) AS Cnt
FROM gfCTE
GROUP BY GF;
```

The first few rows of the result are shown in Figure 7-18. Note that I used only four input variables.

	GF	Cnt
1	1111	1759
2	1112	467
3	1113	12
4	1121	97
5	1122	871
6	1123	450

Figure 7-18. *Count of cases in groups*

Note   You might get a slightly different number of rows in each group. I define the groups for the YearlyIncome column with the NTILE() function, which is not deterministic when it is calculated over an order of a non-unique column.

Some of the groups have far fewer cases than others. Groups with too few cases, such as less than ten, are useless for predictions. That is why I used only four input variables. More input variables would result in more groups with fewer cases. With more data, you could use more input variables and get better predictions. Since there is no training involved, datasets of practically any size can be used.

Let's now split the data into two sets: the training and the test set. I used 30% of the data for testing and 70% for the training set. Note that this is only the split; there is no processing of the data. Listing 7-17 selects 30% of the data randomly for the test set.

Listing 7-17. Preparing the Test Set

```
DROP TABLE IF EXISTS dbo.TMTest;
DROP TABLE IF EXISTS dbo.TMTrain;
GO
-- Test set
SELECT TOP 30 PERCENT
 CAST(
 CASE NumberCarsOwned
  WHEN 0 THEN '1'
  WHEN 1 THEN '1'
  WHEN 2 THEN '2'
  ELSE '3'
 END AS CHAR(1)) +
 CAST(
 CASE CommuteDistance
  WHEN '0-1 Miles' THEN '1'
  WHEN '1-2 Miles' THEN '2'
  WHEN '2-5 Miles' THEN '2'
  WHEN '5-10 Miles' THEN '3'
  WHEN '10+ Miles' THEN '3'
  ELSE '0'
 END AS CHAR(1)) +
 CAST(
  CASE Region
   WHEN 'Europe' THEN '1'
   WHEN 'North America' THEN '2'
```

```
   WHEN 'Pacific' THEN '3'
   ELSE '0'
  END AS CHAR(1)) +
 CAST(NTILE(3) OVER(ORDER BY YearlyIncome) AS CHAR(1))
 AS GF,      -- grouping factor
  CustomerKey,
  NumberCarsOwned, CommuteDistance,
  Region, YearlyIncome AS Income,
  BikeBuyer, 2 AS TrainTest
INTO dbo.TMTest
FROM dbo.vTargetMail
ORDER BY CAST(CRYPT_GEN_RANDOM(4) AS INT);
-- 5546 rows
```

I stored the grouping factor variable in the test set. The other 70% of the data for the training set can be selected, as Listing 7-18 shows.

Listing 7-18. Preparing the Training Set

```
-- Training set
SELECT
 CAST(
 CASE NumberCarsOwned
  WHEN 0 THEN '1'
  WHEN 1 THEN '1'
  WHEN 2 THEN '2'
  ELSE '3'
 END AS CHAR(1)) +
 CAST(
 CASE CommuteDistance
  WHEN '0-1 Miles' THEN '1'
  WHEN '1-2 Miles' THEN '2'
  WHEN '2-5 Miles' THEN '2'
  WHEN '5-10 Miles' THEN '3'
  WHEN '10+ Miles' THEN '3'
  ELSE '0'
 END AS CHAR(1)) +
```

```
CAST(
 CASE Region
  WHEN 'Europe' THEN '1'
  WHEN 'North America' THEN '2'
  WHEN 'Pacific' THEN '3'
  ELSE '0'
 END AS CHAR(1)) +
CAST(NTILE(3) OVER(ORDER BY YearlyIncome) AS CHAR(1))
 AS GF,      -- grouping factor
 CustomerKey,
 NumberCarsOwned, CommuteDistance,
 Region, YearlyIncome AS Income,
 BikeBuyer, 1 AS TrainTest
INTO dbo.TMTrain
FROM dbo.vTargetMail AS v
WHERE NOT EXISTS
 (SELECT * FROM dbo.TMTest AS t
  WHERE v.CustomerKey = t.CustomerKey);
GO
-- 12938 rows
ALTER TABLE dbo.TMTrain ADD CONSTRAINT PK_TMTrain
 PRIMARY KEY CLUSTERED (CustomerKey);
CREATE NONCLUSTERED INDEX NCL_TMTrain_gf
 ON dbo.TMTrain (GF, BikeBuyer);
GO
```

I created an index on the training set using the grouping factor and the bike buyer for the key. This way, I could quickly find the appropriate group for every row of the test set. Let's check the distribution of the target variable in the groups of the grouping factor of the training set, like the following code shows.

```
SELECT GF,
 AVG(1.0 * BikeBuyer) AS Prb,
 COUNT(*) AS Cnt
FROM dbo.TMTrain
GROUP BY GF
ORDER BY GF;
```

I used the test set to simulate the new data. I made the predictions by using the appropriate group from the training set. The test set is a subset of the known data, so the outcome, which is the target variable's value, is known. This way, I could measure the accuracy of the model's predictions.

Performing Predictions with LAM

Using the training set for the predictions is simple. I just need to find the appropriate group for each case from the test set and then calculate the distribution of the target variable in the group and use this for the prediction. Listing 7-19 shows the predictions.

Listing 7-19. Predictions with the Look-Alike Model

```
SELECT t.CustomerKey,
 t.GF,
 t.BikeBuyer,
 i.Prb,
 IIF(i.Prb > 0.5, 1, 0) AS BBPredicted,
 i.Cnt,
 t.NumberCarsOwned, t.CommuteDistance,
 t.Region, t.Income
FROM dbo.TMTest AS t
 OUTER APPLY
  (SELECT AVG(1.0 * BikeBuyer) AS Prb,
    COUNT(*) AS Cnt
   FROM dbo.TMTrain
   WHERE GF = t.GF) AS i
ORDER BY t.CustomerKey;
```

You can see the results of the predictions in Figure 7-19.

	CustomerKey	GF	BikeBuyer	Prb	BBPredicted	Cnt
1	11001	1132	1	0.842105	1	114
2	11009	1332	1	0.602649	1	151
3	11013	3123	0	0.408759	0	137
4	11015	1321	1	0.160493	0	81
5	11017	2331	1	0.454976	0	211
6	11018	2331	1	0.454976	0	211

Figure 7-19. *Predictions with LAM*

Some predictions are correct, and others are incorrect. You can check the quality of the model with the *classification matrix*. The classification matrix (also called a *confusion matrix*) shows the number of positive predictions that are positive (the *true positives*), the number of *false positives* (the prediction was positive, but the actual state is negative), the *true negatives*, and the *false negatives*. Then you calculate the *accuracy* as the sum of the true positives and the true negatives, divided by the total number of cases. Listing 7-20 calculates the classification matrix.

Listing 7-20. Creating the Classification Matrix for the LAM

```
WITH predCTE AS
(
SELECT t.CustomerKey,
 t.GF,
 t.BikeBuyer,
 i.Prb,
 IIF(i.Prb > 0.5, 1, 0) AS BBPredicted,
 i.Cnt
FROM dbo.TMTest AS t
 OUTER APPLY
  (SELECT AVG(1.0 * BikeBuyer) AS Prb,
    COUNT(*) AS Cnt
   FROM dbo.TMTrain
   WHERE GF = t.GF) AS i
)
SELECT BikeBuyer, BBPredicted, COUNT(*) AS Cnt
FROM predCTE
GROUP BY BikeBuyer, BBPredicted;
```

Figure 7-20 is the classification matrix.

	BikeBuyer	BBPredicted	Cnt
1	0	0	1737
2	0	1	1057
3	1	0	980
4	1	1	1772

Figure 7-20. *Classification matrix for the LAM*

The classification matrix shows the accuracy is around 63.3%. This might not sound very good to you; however, please note that I only used four input variables, and there was no training involved. You can use look-alike modeling on datasets of any size.

Naïve Bayes

The final algorithm presented in this chapter is the *Naïve Bayes* (NB) algorithm. The training process for this algorithm is the opposite of decision trees. You calculate the distribution of the input variables in the classes of the target variable to see the associations of pairs of variables. Figure 7-21 shows this calculation graphically. Note that I used the same three input variables that are in Figure 7-17; what changed is the input and the target variable axis.

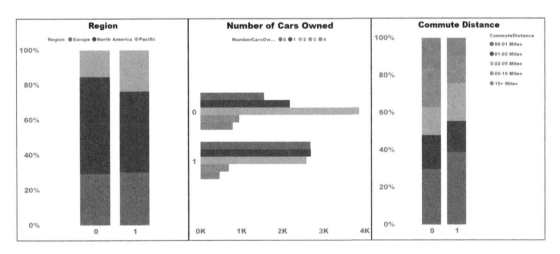

Figure 7-21. *Associations between pairs of variables*

Training an NB model means storing the input variables' probabilities in target variable classes in a table.

Training the NB Model

Let's start by calculating the distribution of a single input variable in a single class of the target variable, as shown in Listing 7-21. Let's use the training set created in the previous section.

Listing 7-21. Calculating the Distribution of an Input Variable for a Single Target State

```
SELECT Region,
 COUNT(*) AS RegCnt,
 1.0 * COUNT(*) / MIN(t1.tot) AS RegPct
FROM dbo.TMTrain
 CROSS JOIN (SELECT COUNT(*)
            FROM dbo.TMTrain
            WHERE BikeBuyer = 1) AS t1(tot)
WHERE BikeBuyer = 1
GROUP BY Region;
```

Figure 7-22 shows this distribution.

	Region	RegCnt	RegPct
1	North America	2952	0.462695924764
2	Pacific	1478	0.231661442006
3	Europe	1950	0.305642633228

Figure 7-22. *Distribution of Region for BikeBuyer equal to one*

Let's use the GROUPING SETS clause to group data over multiple input variables in a single query. I need to separate each state of the target variable and then use the UNION operator to get a single result set. The result set of the query is the model itself. I stored it in a table, as shown in Listing 7-22. Then, the result, or the model, is shown for one combination of input variable values.

Listing 7-22. Creating the NB Model

```
-- Creating the model
DROP TABLE IF EXISTS dbo.TMNB;
SELECT
 1 AS BikeBuyer,
 Region,
 NumberCarsowned,
 CommuteDistance,
 COUNT(*) AS Cnt,
 1.0 * COUNT(*) / MIN(t1.tot) AS Pct
INTO dbo.TMNB
FROM dbo.TMTrain
 CROSS JOIN (SELECT COUNT(*)
             FROM dbo.TMTrain
             WHERE BikeBuyer = 1) AS t1(tot)
WHERE BikeBuyer = 1
GROUP BY
 GROUPING SETS ((Region), (NumberCarsOwned), (CommuteDistance))
UNION ALL
SELECT
 0 AS BikeBuyer,
 Region,
 NumberCarsowned,
 CommuteDistance,
 COUNT(*) AS Cnt,
 1.0 * COUNT(*) / MIN(t0.tot) AS Pct
FROM dbo.TMTrain
 CROSS JOIN (SELECT COUNT(*)
             FROM dbo.TMTrain
             WHERE BikeBuyer = 0) AS t0(tot)
WHERE BikeBuyer = 0
GROUP BY
 GROUPING SETS ((Region), (NumberCarsOwned), (CommuteDistance));
GO
```

```
-- Result for one combination of input values
SELECT *
FROM dbo.TMNB
WHERE Region = N'Pacific'
  OR NumberCarsOwned = 1
  OR CommuteDistance = N'2-5 Miles'
ORDER BY BikeBuyer, Region, NumberCarsOwned, CommuteDistance;
```

Figure 7-23 shows the results, or the contents of the model. For simplicity, I used only three input variables for the NB model.

	BikeBuyer	Region	NumberCarsowned	CommuteDistance	Cnt	Pct
1	0	NULL	NULL	2-5 Miles	978	0.149130832570
2	0	NULL	1	NULL	1554	0.236962488563
3	0	Pacific	NULL	NULL	973	0.148368405001
4	1	NULL	NULL	2-5 Miles	1275	0.199843260188
5	1	NULL	1	NULL	1870	0.293103448275
6	1	Pacific	NULL	NULL	1478	0.231661442006

Figure 7-23. *The content of the NB model*

Now let's use the model for the predictions; however, it is a bit more complicated. More mathematics is involved.

Performing Predictions with NB

Using the table in Figure 7-23, let's calculate each target variable's state for a case where the number of cars owned is 1 and the region is Pacific. For simplicity, I used only these two input variables in Table 7-1, which also includes the Pct column cells.

Table 7-1. *Manual Naïve Bayes Prediction of a Single State*

BikeBuyer	Region	NumberCarsOwned	Pct
0	NULL	1	**A**: 0.236962488563
0	Pacific	NULL	**B**: 0.148368405001
1	NULL	1	**C**: 0.293103448275
1	Pacific	NULL	**D**: 0.231661442006

To calculate the two predicted states, P0 and P1, I used the following formulas.

$$P0 = \frac{A*B}{A*B+C*D}$$

$$P1 = \frac{C*D}{A*B+C*D}$$

The following query manually calculates the predictions for both states using the data from Table 7-1.

```
SELECT
 (0.236962488563 * 0.148368405001) /
 ((0.236962488563 * 0.148368405001) +
  (0.293103448275 * 0.231661442006))
 AS P0,
 (0.293103448275 * 0.231661442006) /
 ((0.236962488563 * 0.148368405001) +
  (0.293103448275 * 0.231661442006))
 AS P1;
```

To make a product from the values in different rows, I would need a special aggregate function, or an *aggregate product*, which does not exist in T-SQL. However, the SUM() function does exist. With the help of logarithm equations, you can use the SUM() function instead of an aggregate product. I used the following equation.

$$\log(a*b) = \log(a) + \log(b)$$

I calculated the sum of the logarithms and then used the EXP() function to get the actual value. Listing 7-23 calculates the probabilities of the positive and the negative state outcomes for the example from Figure 7-23, a case where the region is Pacific, the number of cars owned equals 1, and the commute distance is two to five miles.

Listing 7-23. Predicting for a Single Case

```
DECLARE @nPct AS DECIMAL(18,17),
 @pPct AS DECIMAL(18,17),
 @tPct AS DECIMAL(18,17);
SET @nPct =
(
SELECT EXP(SUM(LOG(Pct))) AS nPct
FROM dbo.TMNB
WHERE BikeBuyer = 0
  AND (Region = N'Pacific'
   OR NumberCarsOwned = 1
   OR CommuteDistance = N'2-5 Miles')
)
SET @pPct =
(
SELECT EXP(SUM(LOG(Pct))) AS pPct
FROM dbo.TMNB
WHERE BikeBuyer = 1
  AND (Region = N'Pacific'
   OR NumberCarsOwned = 1
   OR CommuteDistance = N'2-5 Miles')
)
SET @tPct = @pPct / (@nPct + @pPct);
SELECT @tPct AS PositiveProbability,
 @nPct / (@nPct + @pPct) AS NegativeProbability;
GO
```

Figure 7-24 shows the result. This case is a potential bike buyer.

	PositiveProbability	NegativeProbability
1	0.72129849712159982	0.27870150287840017985

Figure 7-24. *Prediction for a single case*

Let's create a prediction query for a single combination of input variables and return a single predicted state, such as positive prediction only, as shown in Listing 7-24.

Listing 7-24. Single Prediction with a Query

```
-- Prediction query
WITH nPctCTE AS
(
SELECT EXP(SUM(LOG(Pct))) AS nPct
FROM dbo.TMNB
WHERE BikeBuyer = 0
  AND (Region = N'Pacific'
    OR NumberCarsOwned = 1
    OR CommuteDistance = N'2-5 Miles')
),
pPctCTE AS
(
SELECT EXP(SUM(LOG(Pct))) AS pPct
FROM dbo.TMNB
WHERE BikeBuyer = 1
  AND (Region = N'Pacific'
    OR NumberCarsOwned = 1
    OR CommuteDistance = N'2-5 Miles')
)
SELECT pPct / (nPct + pPct) AS tPct
FROM nPctCTE CROSS JOIN pPctCTE;
GO
```

The result is the same as it was for the positive prediction in Figure 7-24. But this query is more appropriate for parameterization.

Listing 7-25 creates a prediction function from Listing 7-24. This function accepts three parameters for the three input variables and returns the probability for the positive state of the target variable.

Listing 7-25. Prediction Function

```
-- Prediction function
CREATE OR ALTER FUNCTION dbo.PredictNB
 (@Region NVARCHAR(50),
  @NumberCarsOwned TINYINT,
  @CommuteDistance NVARCHAR(15))
RETURNS TABLE
AS RETURN
(
WITH nPctCTE AS
(
SELECT EXP(SUM(LOG(Pct))) AS nPct
FROM dbo.TMNB
WHERE BikeBuyer = 0
  AND (Region = @Region
   OR NumberCarsOwned = @NumberCarsOwned
   OR CommuteDistance = @CommuteDistance)
),
pPctCTE AS
(
SELECT EXP(SUM(LOG(Pct))) AS pPct
FROM dbo.TMNB
WHERE BikeBuyer = 1
  AND (Region = @Region
   OR NumberCarsOwned = @NumberCarsOwned
   OR CommuteDistance = @CommuteDistance)
)
SELECT pPct / (nPct + pPct) AS tPct
FROM nPctCTE CROSS JOIN pPctCTE);
GO
```

Now let's use this function for predictions on the complete test set, as shown in Listing 7-26.

Listing 7-26. Full Predictions with the NB Model

```
-- Full prediction on the test set
SELECT t.CustomerKey,
 t.BikeBuyer,
 t.Region,
 t.NumberCarsOwned,
 t.CommuteDistance,
 i.tPct,
 IIF(i.tPct > 0.5, 1, 0) AS BBPredicted
FROM dbo.TMTest AS t
 CROSS APPLY
  dbo.PredictNB(t.Region, t.NumberCarsOwned, t.CommuteDistance) AS i
ORDER BY t.CustomerKey;
```

The partial result is shown in Figure 7-25.

	CustomerKey	BikeBuyer	Region	NumberCarsOwned	CommuteDistance	tPct	BBPredicted
1	11001	1	Pacific	1	0-1 Miles	0.717719525962365	1
2	11009	1	Pacific	1	5-10 Miles	0.579608661197629	1
3	11013	0	North America	3	0-1 Miles	0.45252632491158	0
4	11015	1	North America	1	5-10 Miles	0.423525859955977	0
5	11017	1	Pacific	2	5-10 Miles	0.431690650505846	0
6	11018	1	Pacific	2	5-10 Miles	0.431690650505846	0

Figure 7-25. *Predictions with the Naïve Bayes model*

Again, some predictions are correct, and others are incorrect. Let's calculate the classification matrix again, as shown in Listing 7-27.

Listing 7-27. Creating the Classification Matrix for NB

```
WITH predCTE AS
(
SELECT t.CustomerKey,
 t.BikeBuyer,
 t.Region,
 t.NumberCarsOwned,
 t.CommuteDistance,
 i.tPct,
```

```
 IIF(i.tPct > 0.5, 1, 0) AS BBPredicted
FROM dbo.TMTest AS t
 CROSS APPLY
  dbo.PredictNB(t.Region, t.NumberCarsOwned, t.CommuteDistance) AS i
)
SELECT BikeBuyer, BBPredicted, COUNT(*) AS cnt
FROM predCTE
GROUP BY BikeBuyer, BBPredicted;
```

Figure 7-26 shows this classification matrix.

	BikeBuyer	BBPredicted	Cnt
1	0	0	1803
2	0	1	991
3	1	0	1187
4	1	1	1565

Figure 7-26. *Classification matrix for the NB*

From the classification matrix, the accuracy is around 60.7%. This is slightly worse than the accuracy of look-alike modeling. However, I used only three input variables here. You can use Naïve Bayes algorithms on huge data sets. Some training is done; however, this is efficient training, without loops or recursion.

Conclusion

Another intensive chapter is done. You learned how to implement advanced data mining algorithms with T-SQL. The advantage of using T-SQL is that you can use huge datasets, which might be an important factor when the amount of data you need to analyze grows extremely quickly.

It is time to clean my demo database with the following code.

```
USE AdventureWorksDW2017;
DROP TABLE IF EXISTS dbo.EMA;
DROP TABLE IF EXISTS dbo.tAssocSeqLineItems;
DROP TABLE IF EXISTS dbo.tAssocSeqOrders;
```

```
DROP TABLE IF EXISTS dbo.TMTest;
DROP TABLE IF EXISTS dbo.TMTrain;
DROP FUNCTION dbo.PredictNB;
DROP TABLE IF EXISTS dbo.TMNB;
GO
```

In Chapter 8, you learn about basic text analysis in T-SQL.

CHAPTER 8

Text Mining

This last chapter of the book introduces text mining with T-SQL. Text mining means analysis of texts. Text mining can include semantic search, term extraction, quantitative analysis of words and characters, and more. Data mining algorithms like association rules can be used to get a deeper understanding of the analyzed text. In this chapter, you learn about the following.

- Full-text search

- Statistical semantic search

- Quantitative analysis of letters and words

- Term Extraction

- Association rules for words

I use SQL Server feature called *full-text search* (FTS). FTS enables users to use approximate search when searching for rows with strings that contain the searched term, similar to the approximate search you get in Internet search engines. You can search in string columns and full documents. FTS includes *Statistical Semantic Search* (SSS), a component that enables you to analyze the semantics of the documents. In addition, FTS brings tabular functions that you can use for more advanced text analysis.

Demo Data

You install FTS with SQL Server Setup. FTS can extract text from many different document types. It uses a special *filter* for each document type. To get all the possibilities, install Office 2010 Filter Pack and load the filters. Listing 8-1 checks whether FTS is installed and the filter packs. The commented code shows how to load Office 2010 Filter Pack if you need to do it.

© Dejan Sarka 2021
D. Sarka, *Advanced Analytics with Transact-SQL*, https://doi.org/10.1007/978-1-4842-7173-5_8

Listing 8-1. Checking FTS Installation and Filters

```
-- Check whether Full-Text and Semantic search is installed
SELECT SERVERPROPERTY('IsFullTextInstalled');
-- Check the filters
EXEC sys.sp_help_fulltext_system_components 'filter';
SELECT * FROM sys.fulltext_document_types;
GO
-- Download and install Office 2010 filter pack and SP 2 for the pack
-- Next, load them
/*
EXEC sys.sp_fulltext_service 'load_os_resources', 1;
GO
*/
-- Restart SQL Server
-- Check the filters again
/*
EXEC sys.sp_help_fulltext_system_components 'filter';
SELECT * FROM sys.fulltext_document_types;
GO
*/
-- Office 2010 filters should be installed
```

In the next step, I create demo data for this chapter. Listing 8-2 creates a new database and a single table in it. The table has only a few columns: a primary key and a column for the titles of the documents, and three columns used by the FTS. The docContent column is a binary column for the Word documents I store in this table. The docExcerpt is a short excerpt of the documents stored as a string. And the docType column describes the document type in the docContent column, so the FTS can use the appropriate filter.

Listing 8-2. Database and Table for Demo Data

```
CREATE DATABASE FTSSS;
GO
USE FTSSS;
-- Table for documents for FTSSS
```

```
CREATE TABLE dbo.Documents
(
  Id INT IDENTITY(1,1) NOT NULL,
  Title NVARCHAR(100) NOT NULL,
  docType NCHAR(4) NOT NULL,
  docExcerpt NVARCHAR(1000) NOT NULL,
  docContent VARBINARY(MAX) NOT NULL,
  CONSTRAINT PK_Documents
    PRIMARY KEY CLUSTERED(id)
);
GO
```

I prepared ten Word documents of my old blogs. You can download all ten documents with the code for this book. I load the documents with ten separate INSERT statements. Listing 8-3 shows only the first statement as an example of how the data loaded. Then this demo data is queried.

Listing 8-3. Loading Demo Data Example

```
INSERT INTO dbo.Documents
(Title, docType, docExcerpt, docContent)
SELECT N'Columnstore Indices and Batch Processing',
 N'docx',
 N'You should use a columnstore index on your fact tables,
   putting all columns of a fact table in a columnstore index.
   In addition to fact tables, very large dimensions could benefit
   from columnstore indices as well.
   Do not use columnstore indices for small dimensions. ',
 bulkcolumn
FROM OPENROWSET(BULK 'C:\Apress\Ch08\01_CIBatchProcessing.docx',
                SINGLE_BLOB) AS doc;
SELECT *
FROM dbo.Documents;
```

Figure 8-1 shows the demo data—all ten documents with a short view of the excerpt and content. But you can see the full titles and the key. Please note that I inserted the Transparent Data Encryption document twice, once with ID 5 and once with ID 10. I did it intentionally. The content is completely the same; however, I slightly changed the excerpt.

	Id	Title	docType	docExcerpt	docContent
1	1	Columnstore Indices and Batch Processing	docx	You should use a c...	0x504B03041400...
2	2	Introduction to Data Mining	docx	Using Data Mining ...	0x504B03041400...
3	3	Why Is Bleeding Edge a Different Conference	docx	During high level p...	0x504B03041400...
4	4	Additivity of Measures	docx	Additivity of measu...	0x504B03041400...
5	5	Transparent Data Encryption	docx	Besides sensitive ...	0x504B03041400...
6	6	Introducing JSON	docx	Although XML is a ...	0x504B03041400...
7	7	Always Encrypted	docx	SQL Server 2016 i...	0x504B03041400...
8	8	JSON Functions	docx	You don't just prod...	0x504B03041400...
9	9	Column Encryption	docx	Backup encryption ...	0x504B03041400...
10	10	Transparent Data Encryption	docx	You might want to ...	0x504B03041400...

Figure 8-1. *Demo data*

To use the SSS feature, you need to install the semanticsdb system database. Listing 8-4 explains the process.

Listing 8-4. Installing SSS Database

```
-- Check whether Semantic Language Statistics Database is installed
SELECT *
FROM sys.fulltext_semantic_language_statistics_database;
GO

-- Install Semantic Language Statistics Database
-- Run the SemanticLanguageDatabase.msi from D:\x64\Setup

-- Attach the database
/*
CREATE DATABASE semanticsdb ON
 (FILENAME = 'C:\Program Files\Microsoft Semantic Language Database\
 semanticsdb.mdf'),
 (FILENAME = 'C:\Program Files\Microsoft Semantic Language Database\
 semanticsdb_log.ldf')
 FOR ATTACH;
```

```
GO
*/

-- Register it
/*
EXEC sp_fulltext_semantic_register_language_statistics_db
 @dbname = N'semanticsdb';
GO
*/
/* Check again
SELECT *
FROM sys.fulltext_semantic_language_statistics_database;
GO
*/
```

Let's start with a quick introduction to the capabilities of the FTS and then analyze text with SSS.

Introducing Full-Text Search

To use FTS, you need to create special *full-text indexes*. You store the indexes in a logical container called a *full-text catalog*. Before indexing the data, you can create a list of *stop*words that you do not want to index. There is already a prepared *system stopwords list* with many noisy words in different languages. You can use it, create your own *stoplist* from it, and add your own stopwords.

Listing 8-5 creates a new stoplist from the system stoplist and adds my stopword, *SQL*. I did not want to index this term because my blogs are usually about SQL Server, and practically every blog includes this term, so it does not help with searches.

Listing 8-5. Creating a Stoplist

```
-- Creating the stoplist
CREATE FULLTEXT STOPLIST SQLStopList
FROM SYSTEM STOPLIST;
GO
ALTER FULLTEXT STOPLIST SQLStopList
 ADD 'SQL' LANGUAGE 'English';
```

```
GO
-- Check the Stopwords list
SELECT w.stoplist_id,
 l.name,
 w.stopword,
 w.language
FROM sys.fulltext_stopwords AS w
 INNER JOIN sys.fulltext_stoplists AS l
  ON w.stoplist_id = l.stoplist_id
WHERE language = N'English'
  AND stopword LIKE N'S%';
```

Listing 8-5 checks the content of my stoplist and confirms that there are system stopwords and that my stopwords are included. Figure 8-2 shows the result from the last query.

	stoplist_id	name	stopword	language
1	5	SQLStopList	S	English
2	5	SQLStopList	SQL	English

Figure 8-2. *Partial content of the custom stoplist*

Please note the stoplist ID is 5. This information is needed later. For now, let's create a full-text catalog and index. The syntax is in Listing 8-6.

Listing 8-6. Creating Full-Text Catalog and Index

```
-- Full-text catalog
CREATE FULLTEXT CATALOG DocumentsFtCatalog;
GO
-- Full-text index
CREATE FULLTEXT INDEX ON dbo.Documents
(
  docExcerpt Language 1033,
  docContent TYPE COLUMN doctype
  Language 1033
  STATISTICAL_SEMANTICS
)
```

```
KEY INDEX PK_Documents
ON DocumentsFtCatalog
WITH STOPLIST = SQLStopList,
     CHANGE_TRACKING AUTO;
GO
```

Full-text indexes are populated asynchronously. Therefore, you need to wait a few seconds after creating before using them in your queries. You can check the status of the full-text catalog through the `sys.dm_fts_active_catalogs` dynamic management view, as the following query shows.

```
-- Check the population status
SELECT name, status_description
FROM sys.dm_fts_active_catalogs
WHERE database_id = DB_ID();
```

If the `status_description` column shows the AVAILABLE value, you can start using the indexes from the catalog.

Full-Text Predicates

You search for the rows where a text column includes the searched term with the `CONTAINS` and the `FREETEXT` predicates. With full-text search (FTS), you can search for the following.

- Simple terms: one or more specific words or phrases

- Prefix terms: begins words or phrases

- Generation terms: inflectional forms of words

- Proximity terms: words or phrases close to another word or phrase

- Thesaurus terms: synonyms of a word

- Weighted terms: words or phrases using values with your custom weight

The four queries in Listing 8-7 search for simple terms, with and without logical operators, in the `docExcerpt` character column.

Listing 8-7. Searching for Simple Terms

```
-- Simple query
SELECT Id, Title, docExcerpt
FROM dbo.Documents
WHERE CONTAINS(docExcerpt, N'data');
-- Logical operators - OR
SELECT id, title, docexcerpt
FROM dbo.Documents
WHERE CONTAINS(docexcerpt, N'data OR index');
-- Logical operators - AND NOT
SELECT Id, Title, docExcerpt
FROM dbo.Documents
WHERE CONTAINS(docExcerpt, N'data AND NOT mining');
-- Logical operators - parentheses
SELECT Id, Title, docExcerpt
FROM dbo.Documents
WHERE CONTAINS(docexcerpt, N'data AND (fact OR warehouse)');
```

Instead of showing the result sets, let's look at the number of rows returned by the queries in order of the execution: 8, 9, 7, and 1.

If you enclose the searched term in double apostrophes, you are searching for an exact term. You can use the asterisk (*) character as the wildcard character and search for terms with a specific prefix. With the function inside the CONTAINS predicate, you can search for the strings where the two searched terms are close together. You can define what does this proximity mean, how many words can be maximally between the two searched terms. The examples of these searches are in Listing 8-8. The three queries return 1, 3, and 1 row.

Listing 8-8. Phrase, Prefix, and Proximity Search

```
-- Phrase
SELECT Id, Title, docExcerpt
FROM dbo.Documents
WHERE CONTAINS(docExcerpt, N'"data warehouse"');
-- Prefix
SELECT Id, Title, docExcerpt
```

```
FROM dbo.Documents
WHERE CONTAINS(docExcerpt, N'"add*"');
-- Proximity
SELECT Id, Title, docExcerpt
FROM dbo.Documents
WHERE CONTAINS(docExcerpt, N'NEAR(problem, data)');
```

You can search for the inflectional forms of the words. Listing 8-9 shows two queries: one searches for the simple term while the other for the inflectional forms of the term.

Listing 8-9. Searching for the Inflectional Forms

```
-- The next query does not return any rows
SELECT Id, Title, docExcerpt
FROM dbo.Documents
WHERE CONTAINS(docExcerpt, N'presentation');
-- The next query returns a row
SELECT Id, Title, docExcerpt
FROM dbo.Documents
WHERE CONTAINS(docExcerpt, N'FORMSOF(INFLECTIONAL, presentation)');
GO
```

The result is shown in Figure 8-3. You can see that the word presentation appears in plural. While the first query didn't return any rows, the second query found the row where the word appears in the excerpt, although in the plural form.

Id	Title	docExcerpt

	Id	Title	docExcerpt
1	3	Why Is Bleeding Edge a ...	During high level presentations attendees encoun...

Figure 8-3. *The inflectional form was found*

You can search for synonyms and add your own synonyms to the thesaurus XML file for each supported language. However, this chapter is not about describing the FTS. This section includes a query that uses the FREETEXT predicate, as shown in Listing 8-10.

Listing 8-10. The FREETEXT Predicate

```
SELECT Id, Title, docExcerpt
FROM dbo.Documents
WHERE FREETEXT(docExcerpt, N'data presentation need');
```

The last query returns nine out of ten rows. The FREETEXT predicate is less precise than the CONTAINS predicate. It searches for any form of the searched terms and returns rows where at least one term appears in the searched column. This predicate probably resembles Internet search engines the most.

Full-Text Functions

Full-text predicates help you searching for rows where indexed columns contain the searched string. This is not text mining. The first step into text mining is the full-text function. There are two full-text functions: CONTAINSTABLE and FREETEXTTABLE. They are a counterpart to full-text predicates. With the CONTAINSTABLE function, you get more precise searches than with the FREETEXTTABLE one. Both are tabular functions; you use them in the FROM clause of the query. Both return two columns: [KEY] and [RANK]. The [KEY]column is used for joining the results to the full-text indexed table, to the table's primary key. The [RANK] column is the simplest text mining; the column tells you the document's rank per the searched terms. Listing 8-11 shows a simple example of usage of the CONTAINSTABLE function.

Listing 8-11. Ranking Documents with the CONTAINSTABLE Function

```
SELECT D.Id, D.Title, CT.[RANK], D.docExcerpt
FROM CONTAINSTABLE(dbo.Documents, docExcerpt,
      N'data OR level') AS CT
 INNER JOIN dbo.Documents AS D
  ON CT.[KEY] = D.Id
ORDER BY CT.[RANK] DESC;
```

Figure 8-4 shows the result. You can see which document is determined the best with the searched terms. To calculate the rank, the FTS counts the number of term occurrences, the distance between terms, and much more. Please note in the results that the document that appears twice, the Transparent Data Encryption document, has each time a different rank. Remember that I changed the excerpt a bit, shuffled the words, so the distance between the searched terms is different in the two excerpts.

	Id	Title	RANK	docExcerpt
1	7	Always Encrypted	12	SQL Server 2016 introduced a n...
2	9	Column Encryption	10	Backup encryption encrypts bac...
3	5	Transparent Data Encryption	10	Besides sensitive data, for whic...
4	3	Why Is Bleeding Edge a Different Conference	8	During high level presentations ...
5	10	Transparent Data Encryption	8	You might want to protect all of ...

Figure 8-4. *Ranks of the documents with a simple search*

The rank is always relative to the query. You cannot compare the ranks between queries. Listing 8-12 shows the calculation of rank with the FREETEXTTABLE function.

Listing 8-12.. Ranking Documents with the CONTAINSTABLE Function

```
SELECT D.Id, D.Title, FT.[RANK], D.docExcerpt
FROM FREETEXTTABLE (dbo.Documents, docExcerpt,
      N'data level') AS FT
 INNER JOIN dbo.Documents AS D
  ON FT.[KEY] = D.Id
ORDER BY FT.[RANK] DESC;
```

Figure 8-5 shows the results. The documents are in a different order, and the ranks are completely different. I find the CONTAINSTABLE function much more useful for ranking documents.

	Id	Title	RANK	docExcerpt
1	3	Why Is Bleeding Edge a Different Conference	141	During high level presentation...
2	7	Always Encrypted	133	SQL Server 2016 introduced a...
3	9	Column Encryption	133	Backup encryption encrypts ba...
4	10	Transparent Data Encryption	0	You might want to protect all o...
5	8	JSON Functions	0	You don't just produce JSON f...

Figure 8-5. *Ranks of the documents with the FREETEXTABLE function*

The CONTAINSTABLE function supports all expressions that the CONTAINS predicate supports. In addition, you can define the *weights* of the terms inside the function. Listing 8-13 shows ranking with weighted terms. Note that the sum of the weights is one.

Listing 8-13. Ranking with Weighted Terms

```
SELECT D.Id, D.Title, CT.[RANK], D.docExcerpt
FROM CONTAINSTABLE
      (dbo.Documents, docExcerpt,
       N'ISABOUT(data weight(0.2), level weight(0.8))') AS CT
 INNER JOIN dbo.Documents AS D
  ON CT.[KEY] = D.Id
ORDER BY CT.[RANK] DESC;
```

Figure 8-6 shows the weighted terms' rank. This time, the document that appears twice, the Transparent Data Encryption document, has the same rank. I influenced the rank calculation with weight.

	Id	Title	RANK	docExcerpt
1	7	Always Encrypted	15	SQL Server 2016 introduced a new...
2	9	Column Encryption	14	Backup encryption encrypts backu...
3	3	Why Is Bleeding Edge a Different Conference	9	During high level presentations att...
4	5	Transparent Data Encryption	2	Besides sensitive data, for which y...
5	10	Transparent Data Encryption	2	You might want to protect all of yo...

Figure 8-6. *Documents ranked with weighted terms*

All the queries with the full-text functions returned all documents. If the keywords were not in the documents, the rank was simply zero. But you can limit a search with the CONTAINSTABLE function just like you can with the CONTAINS predicate. Listing 8-14 shows how to use a proximity term.

Listing 8-14. Proximity Term with the CONTAINSTABLE Function

```
SELECT D.Id, D.Title, CT.[RANK]
FROM CONTAINSTABLE (dbo.Documents, docContent,
      N'NEAR((data, row), 30)') AS CT
 INNER JOIN dbo.Documents AS D
  ON CT.[KEY] = D.Id
ORDER BY CT.[RANK] DESC;
```

Figure 8-7 shows the full result. As you can see, only three rows returned.

	Id	Title	RANK
1	1	Columnstore Indices and Batch Processing	4
2	7	Always Encrypted	2
3	2	Introduction to Data Mining	1

Figure 8-7. *Result filtered by the CONTAINSTABLE function*

The next step in text mining uses semantic search functions.

Statistical Semantic Search

You can do three different searches with the SSS functions.

- Search for the semantic key phrases in the documents.

- Rank documents by the semantic similarity to the selected document.

- Find key phrases that are common across two documents.

Let's start with the SEMANTICKEYPHRASETABLE function in Listing 8-15. I searched for the top 100 most important phrases in the list of documents I analyzed.

Listing 8-15. Searching for the Top Semantic Phrases

```
SELECT TOP (100)
 D.Id, D.Title, SKT.keyphrase, SKT.score
FROM SEMANTICKEYPHRASETABLE
      (dbo.Documents, doccontent) AS SKT
 INNER JOIN dbo.Documents AS D
  ON SKT.document_key = D.Id
ORDER BY SKT.score DESC;
```

Figure 8-8 shows the partial result. Again, the phrase *tde* from both Transparent Data Encryption documents has the same score.

	Id	Title	keyphrase	score
1	8	JSON Functions	nvarchar	0.8203845
2	5	Transparent Data Encryption	tde	0.7409303
3	10	Transparent Data Encryption	tde	0.7409303
4	4	Additivity of Measures	additive	0.7003461
5	6	Introducing JSON	xml	0.6957879
6	2	Introduction to Data Mining	undirected	0.6858258
7	9	Column Encryption	symmetric	0.6479381

Figure 8-8. *Top phrases*

Let's order the documents by similarity to document 1, named Columnstore Indices and Batch Processing. Listing 8-16 uses the SEMANTICSIMILARITYTABLE function for this task.

Listing 8-16. Searching for Similar Documents

```
SELECT SST.matched_document_key,
 D.Title, SST.score
FROM SEMANTICSIMILARITYTABLE
     (dbo.Documents, doccontent, 1) AS SST
 INNER JOIN dbo.Documents AS D
  ON SST.matched_document_key = D.Id
ORDER BY SST.score DESC;
```

You can see the results in Figure 8-9. Interestingly, this time the two occurrences of the Transparent Data Encryption document have slightly different scores. I compared the content of the documents, which is completely the same. Therefore, I would expect the same score. The difference between the two scores is very small. Still, this is a good piece of information; the calculation is not completely precise and deterministic.

	matched_document_key	Title	score
1	9	Column Encryption	0.1858189
2	4	Additivity of Measures	0.1794954
3	8	JSON Functions	0.1787596
4	7	Always Encrypted	0.1725576
5	6	Introducing JSON	0.1602058
6	10	Transparent Data Encryption	0.1500729
7	5	Transparent Data Encryption	0.1498567
8	2	Introduction to Data Mining	0.09740368
9	3	Why Is Bleeding Edge a Different Conference	0.07894265

Figure 8-9. *Documents by the similarity to the document one*

Listing 8-17 searches for the semantic key phrases that are common across two documents. Let's compare document 1 with document 4.

Listing 8-17. Searching for the Common Key Phrases

```
SELECT SSDT.keyphrase, SSDT.score
FROM SEMANTICSIMILARITYDETAILSTABLE
      (dbo.Documents, docContent, 1,
        docContent, 4) AS SSDT
WHERE SSDT.keyphrase NOT IN (N'sarka', N'dejan')
ORDER BY SSDT.score DESC;
```

Figure 8-10 shows the result. Note that in the query, I filtered out my first name and last name. My name appears in the metadata of all documents since I am the author, and it doesn't make sense to return my name in the result.

	keyphrase	score
1	metadata	0.2517761
2	dimensions	0.1598883
3	data	0.1207339
4	queries	0.09310722
5	s.p.	0.09123535
6	server	0.07584953

Figure 8-10. *Key phrases that are common across two documents*

Both FTS and SSS support many languages. However, FTS is much more powerful here. Let's check what languages are supported with the queries in Listing 8-18.

Listing 8-18. Querying the Languages Supported by FTS and SSS

```
-- Full Text Languages
SELECT *
FROM sys.fulltext_languages
ORDER BY name;
-- Semantic Search Languages
SELECT *
FROM sys.fulltext_semantic_languages
ORDER BY name;
```

Figure 8-11 shows the result. The first five rows from each query show that not all languages supported by FTS are supported by SSS. FTS supports 53 languages, and SSS supports only 15.

	lcid	name
1	1025	Arabic
2	1093	Bengali (India)
3	1044	Bokmål
4	1046	Brazilian
5	2057	British English

	lcid	name
1	1046	Brazilian
2	2057	British English
3	3076	Chinese (Hong Kong SAR, PRC)
4	5124	Chinese (Macao SAR)
5	4100	Chinese (Singapore)

Figure 8-11. *Languages supported by FTS and SSS*

I've exhausted the out-of-the-box features of FTS and SSS. For further analysis of the text, I needed to create custom queries.

Quantitative Analysis

Quantitative analysis means something was measured, such as letter frequency and word length frequency. In addition, terms that appear at the beginning of the documents might be more meaningful than terms that appear for the first time somewhere in the middle or at the end of the documents you are analyzing.

Analysis of Letters

Letter frequency is a good indicator of the language used in written documents. Letter frequency is particularly effective in identifying whether the writing system in the documents you are analyzing is alphabetic, syllabic, or ideographic. Letter frequency is known for all bigger written languages. For example, you can refer to the article `Letter Frequency`, at `https://en.wikipedia.org/wiki/Letter_frequency`.

Splitting the string into single letters is easy with the auxiliary table of numbers. If you remember, I created a table called `dbo.DateNums` in the AdventureWorksDW2017 database. You can split a string by joining it to the numbers table with a non-equijoin, where the join condition must be lower than or equal to the length of the string. This way, you get a row for every letter in the string. Then the `SUBSTRING()` function extracts the letter on every position, as shown in the following code.

```
SELECT SUBSTRING(a.s, d.n, 1) AS Chr, d.n AS Pos,
  ASCII(SUBSTRING(a.s, d.n, 1)) AS Cod
FROM (SELECT 'TestWord AndAnotherWord' AS s) AS a
 INNER JOIN AdventureWorksDW2017.dbo.DateNums AS d
  ON d.n <= LEN(a.s);
```

I expanded the code to work on a table column, as shown in Listing 8-19.

Listing 8-19. Splitting a String Column to Letters

```
SELECT a.docExcerpt AS Id,
  d.n AS Pos,
  SUBSTRING(a.docExcerpt, d.n, 1) AS Chr,
  UPPER(SUBSTRING(a.docExcerpt, d.n, 1)) AS Chu,
  ASCII(SUBSTRING(a.docExcerpt, d.n, 1)) AS Cod,
  ASCII(UPPER(SUBSTRING(a.docExcerpt, d.n, 1))) AS Cdu
```

```
FROM dbo.Documents AS a
 INNER JOIN AdventureWorksDW2017.dbo.DateNums AS d
  ON d.n <= LEN(a.docExcerpt)
ORDER BY id, pos;
```

You can see the partial result in Figure 8-12. I returned the string I split, the letter, the letter in uppercase, the letter's ASCII code of the uppercase letter's ASCII code.

	Id	Pos	Chr	Chu	Cod	Cdu
1	Additivity of measures is not exactly a data war...	1	A	A	65	65
2	Additivity of measures is not exactly a data war...	2	d	D	100	68
3	Additivity of measures is not exactly a data war...	3	d	D	100	68
4	Additivity of measures is not exactly a data war...	4	i	I	105	73
5	Additivity of measures is not exactly a data war...	5	t	T	116	84

Figure 8-12. *Docexcerpt column characters*

From here, it is simple to do the quantitative analysis. It is just a matter of aggregating data, as shown in Listing 8-20.

Listing 8-20. Analyzing Letters

```
SELECT UPPER(SUBSTRING(a.docExcerpt, d.n, 1)) AS Chu,
  COUNT(*) AS Cnt
FROM dbo.Documents AS a
 INNER JOIN AdventureWorksDW2017.dbo.DateNums AS d
  ON d.n <= LEN(a.docexcerpt)
WHERE ASCII(UPPER(SUBSTRING(a.docExcerpt, d.n, 1))) BETWEEN 65 AND 90
GROUP BY UPPER(SUBSTRING(a.docExcerpt, d.n, 1))
ORDER BY Cnt DESC;
```

The partial results are shown in Figure 8-13.

	Chu	Cnt
1	E	389
2	T	313
3	A	284
4	N	248
5	O	237
6	S	235
7	I	215
8	R	199

Figure 8-13. *Letter frequency*

It looks like the documents are in the English language. At least the first three letters comply with English letter frequency, and other letters are very close to their general position.

Word Length Analysis

To analyze the words, I needed to extract them from the documents. I used the sys.dm_fts_parser() dynamic management function for this task. This function operates on a character column and extracts words in the root form, just as full-text indexing extracts them when creating a full-text index. Listing 8-21 shows the usage of this function.

Listing 8-21. Using sys.dm_fts_parser

```
SELECT display_term, LEN(display_term) AS trlen
FROM dbo.Documents
CROSS APPLY sys.dm_fts_parser('"' + docExcerpt + '"', 1033, 5, 0)
WHERE special_term = N'Exact Match'
  AND LEN(display_term) > 2;
```

You can see the partial results in Figure 8-14. I calculated the length of the extracted terms.

	display_term	trlen
1	columnstore	11
2	index	5
3	fact	4
4	tables	6
5	putting	7
6	columns	7
7	fact	4
8	table	5

Figure 8-14. *Terms and lengths*

Now let's analyze the frequency of the word lengths. Word length distribution is language-specific. You can see the distributions for major languages at www.ravi.io/ language-word-lengths. Listing 8-22 analyzes the word length frequency.

Listing 8-22. Analyzing Word Length Frequency

```
WITH trCTE AS
(
SELECT display_term, LEN(display_term) AS trlen
FROM dbo.Documents
CROSS APPLY sys.dm_fts_parser('"' + docExcerpt + '"', 1033, 5, 0)
WHERE special_term = N'Exact Match'
  AND LEN(display_term) > 2
),
trlCTE AS
(
SELECT trLen,
 COUNT(*) AS Cnt
FROM trCTE
GROUP BY trLen
)
```

```
SELECT trLen, Cnt,
 CAST(ROUND(100. * cnt / SUM(cnt) OVER(), 0) AS INT) AS Pct,
 CAST(REPLICATE('*', ROUND(100. * cnt / SUM(cnt) OVER(), 0))
  AS VARCHAR(50)) AS Hst
FROM trlCTE
ORDER BY trLen;
```

The full result is in Figure 8-15. If you followed the link to the website with word-length frequencies, you might have noticed that in English, the word lengths peak at eight or nine characters. The average word length is 8.23 characters. The words to analyze are a bit shorter than in average English documents. This is because I am not a native English speaker, and there is still room for improvement in my English skills.

	trLen	Cnt	Pct	Hst
1	3	35	10	**********
2	4	61	17	*****************
3	5	39	11	***********
4	6	48	13	*************
5	7	62	17	*****************
6	8	40	11	***********
7	9	27	7	*******
8	10	26	7	*******
9	11	16	4	****
10	12	4	1	*
11	13	1	0	
12	14	2	1	*
13	15	1	0	

Figure 8-15. *Word length frequency in demo documents*

The sys.dm_fts_parser() function returns a column called *occurrence*, which indicates the order of the term in the document. Listing 8-23 finds all the terms and their order.

Listing 8-23. Finding the Order of Terms

```
SELECT Id, Title, display_term, occurrence,
 (LEN(docExcerpt) - LEN(REPLACE(docExcerpt, display_term, '')))
 / LEN(display_term) AS tfIndoc
FROM dbo.Documents
CROSS APPLY sys.dm_fts_parser('"' + docExcerpt + '"', 1033, 5, 0)
WHERE special_term = N'Exact Match'
  AND LEN(display_term) > 2
ORDER BY Id, tfIndoc DESC;
```

Figure 8-16 shows the partial result. For example, the term *columnstore* appears four times in the Columnstore Indices and Batch Processing document, with the first occurrence in fifth place, with four terms before it.

	Id	Title	display_term	occurrence	tfIndoc
1	1	Columnstore Indices and Batch Processing	columns	141	5
2	1	Columnstore Indices and Batch Processing	columnstore	148	4
3	1	Columnstore Indices and Batch Processing	columnstore	5	4
4	1	Columnstore Indices and Batch Processing	columnstore	417	4
5	1	Columnstore Indices and Batch Processing	columnstore	552	4
6	1	Columnstore Indices and Batch Processing	fact	281	3

Figure 8-16. *Terms and their order*

Now let's find the first occurrence of each term in each document with the query in Listing 8-24.

Listing 8-24. Finding First Occurrence of Each Word

```
SELECT Id, MIN(Title) AS Title,
 display_term AS Term, MIN(occurrence) AS firstOccurrence
FROM dbo.Documents
CROSS APPLY sys.dm_fts_parser('"' + docExcerpt + '"', 1033, 5, 0)
WHERE special_term = N'Exact Match'
  AND LEN(display_term) > 2
GROUP BY Id, display_term
ORDER BY Id, firstOccurrence;
```

You can see the result in Figure 8-17.

	Id	Title	Term	firstOccurrence
1	1	Columnstore Indices and Batch Processing	columnstore	5
2	1	Columnstore Indices and Batch Processing	index	6
3	1	Columnstore Indices and Batch Processing	fact	9
4	1	Columnstore Indices and Batch Processing	tables	10
5	1	Columnstore Indices and Batch Processing	putting	139

Figure 8-17. *First occurrences of words*

The first four terms in the results are significant for trying to determine the content of the document pretty well. The order of the terms in a document is not a general indicator of the document's content. Nevertheless, it is common to have a short excerpt at the beginning of technical documents. In such a case, analyzing the terms at the beginning of the documents makes sense. Of course, I have the excerpt stored in a separate column; therefore, the results are intentionally good for showing the usability of the analysis method.

Advanced Analysis of Text

The last section of this chapter covers more advanced text analysis. First is term extraction, meaning extracting all terms. This was done in the previous section; however, I added two different scores to the terms to show which terms are the most important. Then I checked which terms frequently appear together. I used the association rules algorithm with the queries I developed in the previous chapter.

Term Extraction

For scoring the terms, I used *term frequency* (TF), which is a very simple measure of the number of times a term appears. In addition, the *term frequency/inverse document frequency* (TFIDF) score is calculated. The following is the formula for this score.

$$TFIDF = TF * log\left(\frac{N\ docs}{N\ docs\ with\ term}\right)$$

TFIDF lowers the score for terms that appear in many documents and emphasizes frequent terms in a lower number of documents, thus telling us more about those documents. Listing 8-25 calculates both scores.

Listing 8-25. Calculating TF and TFIDF

```
WITH termsCTE AS
(
SELECT Id, display_term AS Term,
 (LEN(docExcerpt) - LEN(REPLACE(docExcerpt, display_term, ''))) /
  LEN(display_term) AS tfindoc
FROM dbo.Documents
CROSS APPLY sys.dm_fts_parser('"' + docExcerpt + '"', 1033, 5, 0)
WHERE special_term = N'Exact Match'
  AND LEN(display_term) > 2
),
tfidfCTE AS
(
SELECT Term, SUM(tfindoc) AS tf,
 COUNT(id) AS df
FROM termsCTE
GROUP BY Term
)
SELECT Term,
 t.tf AS TF,
 t.df AS DF,
 d.nd AS ND,
 1.0 * t.tf * LOG(1.0 * d.nd / t.df) AS TFIDF
FROM tfidfCTE AS t
 CROSS JOIN (SELECT COUNT(DISTINCT id) AS nd
            FROM dbo.Documents) AS d
ORDER BY TFIDF DESC, TF DESC, Term;
```

Figure 8-18 shows the partial result.

	Term	TF	DF	ND	TFIDF
1	encrypt	24	3	10	28.8953473038201
2	key	30	5	10	20.7944154167984
3	xml	36	6	10	18.3897224555613
4	columnstore	16	4	10	14.6606517099865
5	symmetric	8	2	10	12.8755032994728
6	column	13	4	10	11.911779514364
7	document	13	4	10	11.911779514364
8	certificate	9	3	10	10.8357552389325

Figure 8-18. *TF and TFIDF calculated*

TFIDF orders the result, and then by TF. Please note that by the TFIDF score, the term *encrypt* is the most important, although the absolute frequency of this term, the TF score, is lower than it is for the next two terms.

Words Associations

In Chapter 7, I used association rules for market basket analysis. Now I use the same algorithm to analyze my documents. First, the query in Listing 8-26 finds the pairs of the terms. The itemset, in this case, is not two products; it is two terms.

Listing 8-26. Finding Two Word Itemsets

```
WITH termsCTE AS
(
SELECT Id, display_term AS Term, MIN(occurrence) AS firstOccurrence
FROM dbo.Documents
CROSS APPLY sys.dm_fts_parser('"' + docExcerpt + '"', 1033, 5, 0)
WHERE special_term = N'Exact Match'
  AND LEN(display_term) > 2
GROUP BY id, display_term
),
Pairs_CTE AS
(
SELECT t1.Id,
```

```
 t1.Term AS Term1,
 t2.Term2
FROM termsCTE AS t1
 CROSS APPLY
  (SELECT term AS Term2
    FROM termsCTE
    WHERE id = t1.id
      AND term <> t1.term) AS t2
)
SELECT Term1, Term2, COUNT(*) AS Support
FROM Pairs_CTE
GROUP BY Term1, Term2
ORDER BY Support DESC;
```

The result is shown in Figure 8-19.

	Term1	Term2	Support
1	not	data	6
2	data	not	6
3	data	server	4
4	database	server	4
5	encrypted	server	4
6	encryption	server	4
7	server	data	4
8	data	database	4

Figure 8-19. *Two terms itemsets*

It is just one more step to find the association rules and calculate support and confidence, as shown in Listing 8-27.

Listing 8-27. Association Rules for Terms

```
WITH termsCTE AS
(
SELECT id, display_term AS Term, MIN(occurrence) AS firstOccurrence
FROM dbo.Documents
CROSS APPLY sys.dm_fts_parser('"' + docExcerpt + '"', 1033, 5, 0)
WHERE special_term = N'Exact Match'
```

```
  AND LEN(display_term) > 3
GROUP BY Id, display_term
),
Pairs_CTE AS
(
SELECT t1.Id,
 t1.term AS Term1,
 t2.Term2
FROM termsCTE AS t1
 CROSS APPLY
  (SELECT term AS Term2
   FROM termsCTE
   WHERE id = t1.Id
     AND term <> t1.term) AS t2
),
rulesCTE AS
(
SELECT Term1 + N' ---> ' + Term2 AS theRule,
 Term1, Term2, COUNT(*) AS Support
FROM Pairs_CTE
GROUP BY Term1, Term2
),
cntTerm1CTE AS
(
SELECT term AS Term1,
 COUNT(DISTINCT id) AS term1Cnt
FROM termsCTE
GROUP BY term
)
SELECT r.theRule,
 r.Term1,
 r.Term2,
 r.Support,
 CAST(100.0 * r.Support / a.numDocs AS NUMERIC(5, 2)) AS SupportPct,
 CAST(100.0 * r.Support / c.term1Cnt AS NUMERIC(5, 2)) AS Confidence
```

```
FROM rulesCTE AS r
 INNER JOIN cntTerm1CTE AS c
  ON r.Term1 = c.Term1
 CROSS JOIN (SELECT COUNT(DISTINCT id)
             FROM termsCTE) AS a(numDocs)
WHERE r.Support > 1
ORDER BY Support DESC, Confidence DESC, Term1, Term2;
```

Figure 8-20 shows the partial result. Note that I scrolled 15 rows down in the result to show more interesting results.

	theRule	Term1	Term2	Support	SupportPct	Confidence
16	server ---> encryption	server	encryption	4	40.00	100.00
17	data ---> database	data	database	4	40.00	50.00
18	data ---> encrypted	data	encrypted	4	40.00	50.00
19	data ---> encryption	data	encryption	4	40.00	50.00
20	data ---> server	data	server	4	40.00	50.00
21	files ---> data	files	data	3	30.00	100.00

Figure 8-20. *Association rules with support and confidence*

So far, I was analyzing the docExcerpt character column only. Can I analyze the document content in the same way? The answer is yes, but with a different dynamic management function. If you try to parse the document content with the sys.dm_fts_parser() function, you get an error.

Association Rules with Many Items

You can use two dynamic management functions to get the content of the full-text index. The sys.dm_fts_index_keywords() function returns the content for the table; the sys.dm_fts_index_keywords_by_document() function returns the content on the document level, not only on the full table. The following is an example of the second function.

```
SELECT *
FROM sys.dm_fts_index_keywords_by_document(
     DB_ID('FTSSS'), OBJECT_ID('dbo.Documents'));
```

This query returns the term, the document id, the occurrence count, and the column id. Note that I indexed the docExcerpt and docContent columns. Listing 8-28 gets the content of the full-text index together with the document and column names.

Listing 8-28. Getting the Full-Text Index Content

```
SELECT fts.document_id, d.Title,
 fts.column_id, c.name,
 fts.display_term, fts.occurrence_count
FROM sys.dm_fts_index_keywords_by_document(
        DB_ID('FTSSS'), OBJECT_ID('dbo.Documents')) AS fts
 INNER JOIN dbo.Documents AS d
  ON fts.document_id = d.Id
 INNER JOIN sys.columns AS c
  ON c.column_id = fts.column_id AND
     C.object_id = OBJECT_ID('dbo.Documents')
ORDER BY NEWID();  -- shuffle the result
```

Figure 8-21 shows the partial result. Note that the terms come from two columns. There are system terms included, like the term *nn10308* in the sixth row. I exclude those system terms from further analysis.

	document_id	Title	column_id	name	display_term	occurrence_count
1	2	Introduction to Data ...	5	docContent	cubes	1
2	4	Additivity of Measures	5	docContent	discount	1
3	2	Introduction to Data ...	4	docExcerpt	necessity	1
4	7	Always Encrypted	5	docContent	column_master_keys	1
5	8	JSON Functions	5	docContent	converts	1
6	6	Introducing JSON	5	docContent	nn10308	1
7	9	Column Encryption	4	docExcerpt	backups	1
8	2	Introduction to Data ...	5	docContent	introduction	1

Figure 8-21. *The content of the full-text index*

There are many more terms in the document content than in the document excerpt. Any term can be in an itemset with another term; therefore, the number of itemsets can grow quadratically with the number of terms. So, you need to be careful with performance.

The first stem filters anything that is not useful for further analysis: to browse a single column, to filter the terms that are too short, such as less than four characters long, to filter all system terms that are not useful for the analysis, and, if needed, to return the most frequent terms only. In addition, a common table expression's performance becomes unacceptable if it returns too big a dataset. SQL Server cannot estimate correctly the number of rows returned; the cardinality estimator makes a rule of thumb estimation. Because of that, the query optimizer many times cannot prepare the optimal execution plan. The solution is to store the partial results of the common table expressions in temporary or permanent tables. SQL Server can calculate the cardinality properly, and you can index the tables. Listing 8-29 creates a temporary table with the rules only.

Listing 8-29. Storing the Rules

```
WITH termsCTE AS
(
SELECT document_id AS Id,  display_term AS Term
FROM sys.dm_fts_index_keywords_by_document(
     DB_ID('FTSSS'), OBJECT_ID('dbo.Documents'))
WHERE display_term <> N'END OF FILE'
  AND LEN(display_term) > 3
  AND column_id = 5
  AND display_term NOT LIKE '%[0-9]%'
  AND display_term NOT LIKE '%[.]%'
  AND display_term NOT IN (N'dejan', N'sarka')
),
Pairs_CTE AS
(
SELECT t1.Id,
 t1.term AS Term1,
 t2.Term2
FROM termsCTE AS t1
 CROSS APPLY
  (SELECT term AS Term2
   FROM termsCTE
```

```
   WHERE id = t1.Id
      AND term <> t1.Term) AS t2
),
rulesCTE AS
(
SELECT Term1 + N' ---> ' + Term2 AS theRule,
 Term1, Term2, COUNT(*) AS Support
FROM Pairs_CTE
GROUP BY Term1, Term2
)
SELECT *
INTO #rules
FROM rulesCTE;
```

With all the filters implemented, there are still more than 500,000 rules in the temporary table. The next step is to store the frequency of the single items to calculate the confidence, which is done in Listing 8-30.

Listing 8-30. Calculating Frequency of Single Items

```
WITH termsCTE AS
(
SELECT document_id AS Id,  display_term AS Term
FROM sys.dm_fts_index_keywords_by_document(
      DB_ID('FTSSS'), OBJECT_ID('dbo.Documents'))
WHERE display_term <> N'END OF FILE'
  AND LEN(display_term) > 3
  AND column_id = 5
  AND display_term NOT LIKE '%[0-9]%'
  AND display_term NOT LIKE '%[.]%'
  AND display_term NOT IN (N'dejan', N'sarka')
),
cntTerm1CTE AS
(
SELECT term AS term1,
 COUNT(DISTINCT id) AS term1Cnt
```

```
FROM termsCTE
GROUP BY term
)
SELECT *
INTO #cntTerm1
FROM cntTerm1CTE;
```

The final query is shown in Listing 8-31.

Listing 8-31. Association Rules for Terms from the Documents' Content

```
SELECT r.theRule,
 r.Term1,
 r.Term2,
 r.Support,
 CAST(100.0 * r.Support / a.numDocs AS NUMERIC(5, 2))
  AS SupportPct,
 CAST(100.0 * r.Support / c.term1Cnt AS NUMERIC(5, 2))
  AS Confidence
FROM #rules AS r
 INNER JOIN #cntTerm1 AS c
  ON r.Term1 = c.Term1
 CROSS JOIN (SELECT COUNT(DISTINCT id)
              FROM dbo.Documents) AS a(numDocs)
WHERE r.Support > 1
  AND r.Support / c.term1Cnt < 1
ORDER BY Support DESC, Confidence DESC, Term1, Term2;
```

Figure 8-22 shows the partial result.

	theRule	Term1	Term2	Support	SupportPct	Confidence
1	data ---> used	data	used	9	90.00	90.00
2	dotm ---> used	dotm	used	9	90.00	90.00
3	normal ---> used	normal	used	9	90.00	90.00
4	used ---> first	used	first	8	80.00	88.89
5	used ---> simple	used	simple	8	80.00	88.89
6	used ---> using	used	using	8	80.00	88.89
7	data ---> first	data	first	8	80.00	80.00
8	data ---> however	data	however	8	80.00	80.00

Figure 8-22. *The rules with the support and the confidence*

This is the last result in this book.

Conclusion

You have come to the end of the book. In this chapter, you saw how to analyze text with pure T-SQL. You can do analyses in multiple languages, analyze string columns, and analyze full documents.

For the last time, I clean my SQL Server instance with the following code.

```
-- Clean up
DROP TABLE IF EXISTS #cntTerm1;
DROP TABLE IF EXISTS #rules;
GO
USE master;
DROP DATABASE FTSSS;
GO
```

I hope you found this book valuable. If you are a database developer using SQL Server or Azure SQL Database, you have seen that you do not need to learn a new language to perform an analysis. In addition, if you use other languages for statistical and data science analyses, you can benefit from using T-SQL for these tasks. Often, the performance of T-SQL is magnitudes better.

Index

A, B

Aggregating strings, 82–85

Analytical/business intelligence (BI) project, 69

Application and system times
    auxiliary table, 139
    dbo.DateNums table, 141
    demo data, 139–144
    demo sales table creation, 141
    granularity, 139
    inclusion constraint, 138, 139
    sales table, 142, 143
    time constraints, 137

Association rules algorithm
    condition/consequence, 232
    confidence, 234
    descending order, 237
    interpretation, 235
    itemsets, 225, 229–231, 237–239
    lift process, 235, 236
    line number, 239
    market basket analysis, 225
    negative associations, 226–229
    popular products, 225
    results, 232
    sequences, 240
    support/confidence, 233

Associations (pairs of variables)
    ANOVA results, 62
    continuous variable
        causation, 42
        correlation coefficient, 38–41
        covariance, 35–37
        interpretation, 41–43
        spurious correlation, 42
        testing, 58
    descriptive statistics, 56
    discrete variables
        chi-squared value, 51–55
        chi-squared test, 43
        contingency tables, 43–51
        crosstabulations, 43
        graphical analysis, 43
        grouping data, 47
    distinct numeric values, 31
    fictious observed probabilities, 32
    F-ratio distributions, 61
    integration, 63–66
    null hypothesis, 32
    possibilities, 31
    significance, 31
    standard normal distribution, 33, 34, 64
    transmission, 56
    testing, 57, 58
    trapezoidal rile, 63
    variance, 58–63

C

Chi-squared distribution, 51–55

Conditional entropy, 130–133

Z

Printed in the United States
by Baker & Taylor Publisher Services